Microbes in Acti

MICROBES IN ACTION

A Laboratory Manual
of Microbiology

FOURTH EDITION

Harry W. Seeley, Jr.
CORNELL UNIVERSITY

Paul J. VanDemark
Late of CORNELL UNIVERSITY

John J. Lee
CITY COLLEGE OF THE CITY UNIVERSITY OF NEW YORK

W. H. Freeman and Company
New York

To the late Paul J. VanDemark, a devoted and talented teacher and researcher, whose spirit, ideas, and dedication live on

Cover Photos: Scanning electron microscopy of the microbial community growing on the surface of *Halophila stipulacea*, an underwater seagrass, at 25 meters. **Top:** The oval diatom *Cocconeis*, at the lower left, and various bacteria cover the leaf. The fuzzy rhizopodial network of a foraminiferan, a shelled protist that is digesting them, covers part of the diatom and various bacteria (20,000×). **Middle:** The profile of individual cells on the surface of *Halophila* (10,000×). **Bottom:** "Protozoan's-eye" view of a part of the leaf that has unicellular hairs extending from the surface. One can imagine a protozoan herbivore grazing in this "jungle" of pennate diatoms and bacteria (10,000×). (Photos by John J. Lee)

The development of some of the new exercises in this manual was aided by the support of NFS Instrumentation and Laboratory Improvement Program grant DIR-885287. Any opinions or recommendations that arise are those of the authors and do not necessarily reflect the views of the National Science Foundation.

ISBN 0-7167-2100-7

Printed in the United States of America

© 1962, 1972, 1981, 1991 by W. H. Freeman and Company

Eighth printing, 2001

CONTENTS

P R E F A C E

From this late twentieth-century vantage point, the study of microorganisms appears central to the advance of many aspects of molecular biology, genetics, molecular evolution, molecular ecology, biotechnology, and other related fields. The interval between publication of the last edition of *Microbes in Action* and this one has seen an explosion of interest in the diversity of microbes, including the unicellular eukaryotic protists. It was in this latter context that the late Paul VanDemark asked me to join the *Microbes in Action* team.

The choices of the new areas and techniques of microbiology that should be included in a basic laboratory manual are not easy to make. As seasoned classroom instructors, we recognize that there should be some overlap with courses in genetics, cell and molecular biology, and immunology, but we also recognize that a single course cannot cover the entire range of related courses. It seems reasonable to design a laboratory course that can prepare students for further work in these fields.

In drawing up the aims of this revision, I have tried to remain faithful to the principles of the original authors. Their primary objectives were to provide basic knowledge in understanding microorganisms and in handling them effectively and safely. Their approach was ecological, attempting to give the student a deep appreciation of the natural relationships of microorganisms to their habitats.

One of my goals was to stimulate students to think more about the microbes in the world with which they have daily contact. The samples they bring to the laboratory for various exercises should pique the intellectual curiosity of most students and help to make the laboratory experience a more personal one.

As in prior editions, there are more classroom exercises than can be reasonably included in a 14-week semester. Instructors can pick and choose among them to meet their educational goals. In some cases, students can choose particular exercises for independent projects. Some exercises that were included in earlier editions have been dropped to keep the manual to a reasonable size. All of the new exercises have been student-tested either at Cornell University or at City College of the City University of New York (CUNY). Continuing the tradition set by Harry Seeley and Paul Van-Demark, I would be grateful to receive any criticisms of this edition and suggestions for the next.

I would like to express my thanks to the following people, whose comments as reviewers were so helpful in preparing this new edition:

Allan T. Andrew, Indiana University of Pennsylvania
Russell G. Barnekow, Jr., Southwest Missouri State University
Jack M. Bostrack, University of Wisconsin—River Falls
H. Tak Cheung, Illinois State University
Mary Lynne Perille Collins, University of Wisconsin
Richard J. Ellis, Bucknell University
E. W. Frampton, Northern Illinois University
David Kafkewitz, Rutgers University—Newark
Gerry Luginbuhl, North Carolina State University
Richard L. Myers, Southwest Missouri State University
Loy Dean Pike, Indiana University—South Bend
Mary Lou Potter, Louisiana State University
Jerry L. Tammen, Rochester Community College

Linda Treeful, University of Wisconsin — River Falls
Marilyn J. Tufte, University of Wisconsin — Platteville
A. Thomas Weber, University of Nebraska — Omaha

I am grateful to the many instructors and students at Cornell and City College who, over the years, have been the "guinea pigs" for new laboratory exercises. I am particularly indebted to Norman Schwartz and Carmine Mastropaolo at City College and Carole Rehkugler, Ereign Seacord, and Dave Hinman at Cornell for their suggestions, which led to the refinement of the new exercises. Stephen Zinder at Cornell was helpful in revising several exercises. Monica J. Lee also gave valuable advice on many technical aspects of this revision. I appreciate Harry Seeley's review of the revised and new material in this fourth edition. Additional thanks go to Shirley Rosenblatt for her typing of the manuscript, and to my wife, Judith G. Lee, for her assistance in its preparation.

John J. Lee
January 1990

SUGGESTIONS AND REGULATIONS

General Suggestions and Information

1. Many of the stains used in microbiology laboratories do not come out if they get onto your clothing. It is sensible to wear a lab coat, smock, or apron.
2. Before each laboratory period, read the exercises to be done and plan your work carefully.
3. Do not begin work until you have received your instructions. Every laboratory meeting will begin with a short instruction period. Do not hesitate to ask questions if the procedures in the laboratory manual or the directions of your instructor are not clear to you.
4. Record all observations at the time they are made. Laboratory examinations will cover the information given by your laboratory instructor, the information given in this manual, and your own observations and conclusions.
5. Answer the questions on the Report sheet after each exercise.

Laboratory Regulations

It is important for you to develop a positive and respectful attitude toward microorganisms and your work with them. Casualness in technique can cause contamination of your work and infection of yourself and others when you are working with potential pathogens.

1. Never trust the student who worked at your laboratory bench before you, but always respect the student who comes after you. Begin every laboratory period by washing down your bench and end it by washing it down again.

2. Keep your desk free of nonessential materials.
3. Never pipet materials by mouth. Use the bulbs or other pipetting devices provided.
4. Do not smoke, eat, or apply cosmetics in the microbiology laboratory.
5. Wash your hands after every laboratory period.
6. Place all contaminated glassware, plasticware, and paper in the appropriate containers provided. At some universities there are four separate places for waste: (1) autoclave for plastic disposables, (2) autoclave for reusable glassware, (3) containers for used pipets, and (4) containers for noncontaminated waste. Never lay contaminated materials on your desk. Remember to sterilize inoculating needles and loops after you use them and before you lay them down on your desk. Remember to remove all your labels from the reusable glassware before you place it in the recycling bin. If everyone helps, it makes it much easier for the dishwasher.
7. Discard all paper and plastic wrapping from disposable materials as soon as practical. There is always the danger of fire if there is an accumulation of combustibles.
8. Do not wear loose hanging clothes or scarves that could catch fire or knock over reagents.
9. Some of the reagents used in this manual are hazardous (e.g., Ame's test). Make sure you wear disposable gloves and follow directions very carefully. Make certain you understand the precautions pointed out by your instructor.
10. Report all accidents or spills to the instructor, no matter how minor. Such accidents will not count against your grade. Do you know what to do if you drop a culture while transferring it? What do you do in case of an alcohol fire?

Microscopes

Most microorganisms are too small to be visible to the unaided eye, so microbiologists employ a variety of microscopes to study them. There are five types of light microscopes—**bright-field, phase-contrast, fluorescence, dark-field,** and **interference**—and two types of **electron microscopes**, the transmission electron microscope (**TEM**) and the scanning electron microscope (**SEM**). Each of these instruments has advantages for particular uses, and mastery of one or more of them is part of the training of every microbiologist.

BRIGHT-FIELD LIGHT MICROSCOPE

The most widely used optical instrument is the bright-field microscope (Figure I-1), which requires the least amount of training to use and can produce a useful magnification of 1000×. A greater magnification than this gives a fuzzy image because of a physical phenomenon known as *resolution* or resolving power. Resolution is the ability of an instrument to optically separate two very close objects in the field so that they can be distinguished individually and not as a single object. In theory, a light microscope can resolve objects as small as 0.2 μm, but in practice resolution varies from 0.3 to 5.0 μm.

$$\text{resolving power} = \frac{\text{wavelength of light used}}{\underset{\text{of objective}}{\text{numerical aperture}} + \underset{\text{of condenser}}{\text{numerical aperture}}}$$

The **numerical aperture** (NA) is a measure of size of the cone of light that can be gathered by the objective. If you look at the objective lenses of a microscope, you will usually find the numerical aperture engraved on them. As magnification is increased, one must design lenses with increased numerical aperture. Student microscope 100× objectives usually have an NA of 1.25. More expensive research microscope lenses can have an NA of 1.3 or 1.4. Numerical aperture is a function of the diffraction pattern formed at the aperture of the lens. At higher magnifications (60 to 100×), immersion oil is placed between the specimen (or cover glass) and the immersion lens to decrease the diffraction of light as it passes into the objective (Figure I-2).

The most commonly used objective lenses on a bright-field microscope are 10×, 40 or 44×, and 95 or 100×. Used with a 10× eyepiece, total magnifications of 100, 440, and 950 diameters are common. **Working distance**, the space between the objective and the specimen, decreases as the focal distance (which is related inversely to magnification) decreases (Figure I-2).

If you examine the formula for resolving power closely, you will see that the numerical aperture of the condenser is an important factor. High-resolution requires that parallel rays of light enter the condenser and that the condenser be focused on the object. In expensive microscopes, this is accomplished by placing an iris diaphragm on the light source (field diaphragm) and focusing the inner leaves of this diaphragm in the object field. This is called **Köhler illumination** (Figure I-3). In some less costly microscopes,

A

Figure I-1 Bright-field microscope showing: (*A*) location of principal components and (*B*) diagram of optical path. (Courtesy of Olympus Optical Co.)

B

Figure I-2 The principle of decreasing the refraction of light by oil immersion. *A*: Five rays from *P* into the object through the coverslip into the air space between the coverslip and the lens. Only rays 1 and 2 can enter the objective. Rays 4 and 5 are totally reflected. *B*: The air space is replaced by oil of the same refractive index as glass. More rays from *P* now pass straight through without deviation. The NA is thus increased by the factor *n*, the refractive index of oil. *C*: We see a comparison of the angular apertures of a high-dry and an oil-immersion objective.

the condenser is focused on the edges of the lamp housing. Light intensity is adjusted by means of a variable transformer on the lamp housing or by neutral density filters. The **substage diaphragm** is never used to control light intensity but to control the size of the light cone entering the objective. If the diaphragm is closed too far, it reduces the numerical aperture of the condenser. In expensive microscopes, a telescope or an intermediate lens is placed in the optical system to adjust the substage diaphragm so that its image is projected into the periphery of the back of the objective.

The size of the cone of light passing into a microscope differs with objectives. As the magnification of the objective lens increases, the working distance decreases (Figure I-2) and the angle of aperture of the objective increases. Therefore, with increasing magnifi-

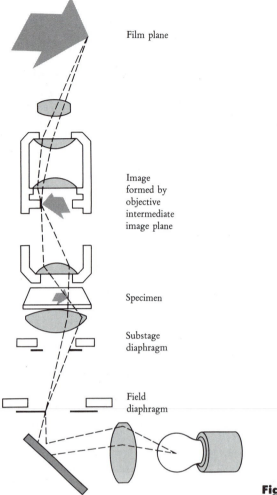

Film plane

Image formed by objective intermediate image plane

Specimen

Substage diaphragm

Field diaphragm

Figure I-3 Köhler illumination.

cation a larger cone of light must enter the objective. When the low-power and high-dry objectives are used, the iris diaphragm is not opened fully because at these magnifications the definition and detail are clearest if the light is not too intense. The working distance with the oil-immersion objective is the shortest, so the iris diaphragm is opened more (Figure I-4).

SPECIAL PRECAUTIONS FOR USING MICROSCOPES

To keep the microscope and lens systems clean:

1. Never touch the lenses. If the lenses become dirty, blow the dust off with an air blower and wipe them gently with lens paper.
2. Never leave a slide on the microscope when it is not in use.
3. Always remove oil from the oil-immersion objective after you use it. If oil should get on a lower-power objective, wipe it off immediately with lens paper. If oil becomes dried or hardened on a lens, you can remove it with lens paper that is lightly moistened with alcohol or *xylol*. *Caution:* Too much xylol will dissolve the cement that holds the lens. A lens paper moistened lightly with tissue culture detergent also cleans greasy and oiled lenses very well without leaving any film.

Figure I-4 Relationship between working distance of objective lens and adjustment of iris diaphragm. The shorter the working distance, the more the diaphragm must be opened.

4. Keep the stage of the microscope clean and dry. If any liquids are spilled, dry the stage with cheesecloth. If oil should get on the stage of the microscope, moisten a piece of cheesecloth with xylol, clean the stage, and wipe it dry.

5. Do not tilt the microscope when working with the oil-immersion system. The oil may flow under the mechanical stage system where it will be difficult to remove, or it may drip onto the substage condenser and harden there.

6. When the microscope is not in use, keep it covered and in the microscope compartment.

To avoid breaking the microscope:

1. Never force any knob on the microscope. All adjustment devices should work easily. If something does not work correctly, do not attempt to fix it yourself; immediately call your instructor.
2. Never allow an objective lens to touch the cover glass or the slide.
3. Never lower the body tube with the coarse adjustment while you are looking through the microscope.
4. Never exchange the objectives or oculars of different microscopes; never, under any circumstance, remove the front lenses from objectives.

USE OF THE TRANSMITTED LIGHT MICROSCOPE

Simple Student Microscope (microscopes without field diaphragms)

1. Place a slide on the stage, specimen side up, and center the section to be examined as accurately as possible over the hole in the center of the stage.
2. Adjust the light source until light passes through the specimen. You may need and be able to adjust the light intensity again once a specimen is in focus. With the low-power objective in position, lower the body tube by means of the coarse adjustment until the objective is about ¾ inch from the side. *Remember:* Never lower the body tube while looking through the ocular. Always watch the objective lens from the side while lowering it toward the slide; then look through the ocular and focus by moving the objective lens *away* from the slide.
3. Look through the eyepiece and slowly raise the objective with the coarse adjustment until the specimen is in approximate focus. Never focus downward while looking through the eyepiece. Bring the specimen into sharp focus with the fine adjustment. Raise the substage condenser all the way up and open the iris diaphragm until the edge of the diaphragm just disappears from view.
4. Now you are ready to shift to the high-dry objective. Examine the specimen with the low-power objective and make certain that the portion of the specimen you want to view is exactly centered in the field. This is necessary because the diameter of the microscopic field under high magnification is much smaller than it is under low magnification. Rotate the nosepiece until the objective clicks into place. *Caution:* Do not touch the lenses of the objectives.
5. The object should be nearly in focus. Look through the eyepiece and slowly adjust the focus (away from the slide!) with the coarse adjustment until the specimen comes into approximate focus. Then bring the image into accurate focus with the fine adjustment.
6. Focusing the oil-immersion objective requires more care than focusing the other objectives, but the procedure is essentially the same. First use the low-power objective to locate the portion of the specimen to be examined. Be extremely careful that this portion is in the exact center of the low-power field because the field diameter is much smaller with the oil-immersion objective than with either of the other objectives. Raise the body tube and rotate the nosepiece until the oil-immersion objective clicks into position. Place a drop of immersion oil on the portion of the slide that you want to view. Watching the objective from the side, carefully lower it into the oil. *Caution:* Do not allow the objective to touch the slide. Look through the ocular and slowly focus upward with the fine adjustment until the image appears. The image will come into view quickly because the working distance of the oil-immersion objective is relatively short. After the image appears, obtain the clearest possible image by critically focusing with the fine adjustment and by adjusting the light intensity and the iris diaphragm to obtain optimum illumination. If you have difficulty in bringing the image into

view, move the stage adjustment continually back and forth while focusing. The motion makes it easier to perceive the image as it comes into focus.

After every use of the oil-immersion objective, clean the oil from the objective lens with lens paper.

Advanced-Student/Research-Grade Microscope (microscopes with field diaphragms [Figure I-5]).

1. Clean microscope surfaces. Blow dust off lenses (ocular, objective, condenser) and top of lamp mount. Wipe with lens paper if necessary.
2. Turn on light.
3. Flip in the neutral density filter if there is one.
4. Center the lamp mount at the base of the microscope by means of the two screws

Light path selector knob

The knob can be operated in 3 positions to deflect the light as desired.

Specimen holder

Stage clamping screw

Loosen the screw and the stage can be rotated.

Aperture iris diaphragm ring

Condenser centering knobs

Voltmeter

Main switch

Field diaphragm

Tension adjustment ring

Clockwise rotation increases coarse adjustment tension.

Coarse adjustment knob

Coarse adjustment range: 26 mm

Fine adjustment knob

Graduated in increments of 2μ.

Sliding voltage control lever

Voltage lowers as the lever is pulled toward the microscope front.

Figure I-5 Advanced student and research microscope. (Courtesy of Olympus Optical Co.)

in the lamp housing. The centering is accomplished by visually aligning the mount and the housing. When they are approximately centered, the same amount of thread will be exposed on both centering screws.

5. Pull out the phase-ring centering screws in the condenser (if they are present).

6. Rotate the turret ring to the "no color spot" or 0 setting.

7. Place a slide on the stage and focus.

8. Close the field diaphragm almost completely.

9. Focus the condenser by means of the large knob under the stage (Figure I-6). When the condenser is in focus you will see the sharp outline of the leaves of the field diaphragm. The illumination obtained by this method is called Köhler illumination (Figure I-3). There should be a blue or red interference halo at the edges of the leaves.

10. Locate condenser centering screws on the bottom part of the condenser mount and use them to center the field diaphragm (Figure I-7).

11. Open the field diaphragm until the leaves touch the periphery (edge) of the field. If you have not centered the condenser perfectly (some leaves touch the periphery and others do not), recenter it.

12. Open the field diaphragm just beyond the field of view. *Caution:* Do not touch the field diaphragm again; Do not touch the condenser adjustment knob again!

13. Swing in the aperture viewing prism by means of the lever that is located above the arm between the nosepiece and the body tube. It is focused by means of the sliding knob located within the lever. *Note:* Some microscope manufacturers use an intermediate projection lens or a telescope to do the examination described in Step 14.

14. Locate the substage iris diaphragm lever. Use this lever to close the iris until it just appears at the outer edge of the objective aperture. The substage iris diaphragm must be adjusted for each objective; if you change objectives, you must readjust the substage diaphragm. As you go to lenses with higher NA, you must open the diaphragm more. *Note:* Light intensity is controlled by use of the neutral density

Figure I-6   Substage condenser assembly and field diaphragm.

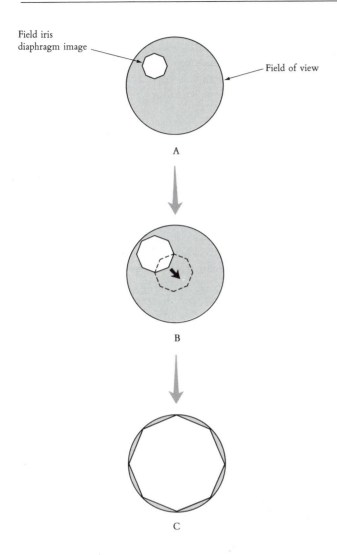

Field iris
diaphragm image

Field of view

A

B

C

Figure I-7 Focusing and centering a field diaphragm. *A*: A focused but eccentric field diaphragm. *B*: A more opened but eccentric field diaphragm. *C*: A properly centered and opened field diaphragm.

filter and the variable transformer. The density filter is not usually used when the phase 40×, or the 100× objectives are in place.

15. Flip out the aperture viewing prism.

1. *Examining Stained Cells*

Because of their minute size, bacteria are not usually studied with the low-power or high-dry objectives; instead, they are stained and observed with the oil-immersion objective. This exercise will give you experience in the use of the oil-immersion objective. It will also serve as a comparative study of the morphology of these stained cells, not only bacteria but also other typical microbes.

PROCEDURE

1. Before starting this exercise, review the instructions for the use of the oil-immersion objective.
2. Then, under the oil-immersion objective, examine stained slides of the following cells:
   (a) *Euglena gracilis*, a protist
   (b) *Paramecium aurelia*, a protist
   (c) *Staphylococcus aureus*
   (d) *Bacillus anthracis*
   (e) *Streptococcus pyogenes*
   (f) *Saccharomyces cerevisiae*, a yeast
   (g) *Salmonella typhosa*
   (h) *Rhodospirillum rubrum*
   (i) *Actinomyces bovis*
3. Observe the sizes and shapes of these organisms and any internal structures that are visible by viewing several fields that contain several isolated cells.

4. On the report sheet, draw each organism you observe, including any visible structures. Make your drawings sufficiently large; for example, draw a coccus as a small circle or ellipsoid, not as a dot.

QUESTIONS

1. How would a stained preparation appear under the oil-immersion objective without the oil between the slide and the objective lens? Why? If you don't know, set up a preparation this way and look at it.
2. What structural differences among the types of cells could you observe in this exercise? Could you detect the presence or absence of structures in the bacterial cells similar to those seen in the other cells? Why?

Examining Stained Cells

Name _____

Desk No. _____

Make drawings of your observations. Make your drawings large enough to illustrate significant features, and indicate relative size.

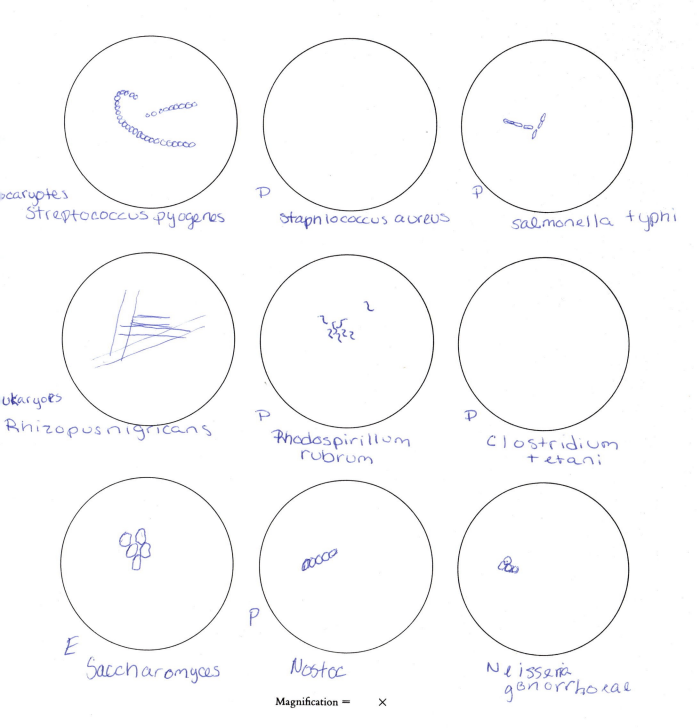

ocaryotes
Streptococcus pyogenes

P
Staphlococcus aureus

P
Salmonella typhi

ukaryoes
Rhizopus nigricans

P
Rhodospirillum rubrum

P
Clostridium tetani

E
Saccharomyces

P
Nostoc

Neisseria gonorrhoeae

Magnification = ___ ×

2. *Microscopic Measurement*

An **ocular micrometer** is a disc of glass on which equally spaced lines are etched. When this disc is placed in the ocular, the ruled lines are superimposed onto the microscope field. The scale of these lines relative to the microscope field (which changes with magnification) can be calibrated by superimposing them onto a **stage micrometer,** which is placed in the focal plane of the microscopic field and on which parallel lines exactly 0.01 mm (10 μm) apart are etched. By determining how many units of the ocular micrometer superimpose a known distance on the stage micrometer, you can calculate the exact distance that each division of the ocular micrometer measures on the microscopic field. Once calibrated, the ocular micrometer can be used to determine the sizes of various microscopic objects. However, when you change objectives or microscopes, you must recalibrate the ocular micrometer.

In this exercise you will calibrate the ocular micrometer and then use it to measure the sizes of typical algae, protozoa, yeasts, and bacteria.

PROCEDURE

1. Remove the ocular lens and insert the ocular micrometer on the circular shelf. Replace the ocular lens and mount in the microscope and observe.
2. Mount the stage micrometer on the microscope stage, and, after centering the scale of the stage micrometer with the low-power objective, observe the micrometer under the oil-immersion objective.
3. Rotate the ocular until you superimpose the lines of the ocular micrometer on those of the stage micrometer. With the lines of the two micrometers coinciding at one end of the field, count the spaces of each micrometer to the point at which the lines of the micrometers coincide again.

   Since each space of the stage micrometer equals 10 μm (micrometer) in distance, and since you know how many ocular divisions are equivalent to one stage division, you can

A	B	C	D
Ocular micrometer in microscope eyepiece; divisions arbitrary.	Stage micrometer. Distances absolute; each division equals 10 micrometers.	The ocular micrometer superimposed on the stage micrometer. In this lens system, what is the absolute length in micrometers covered by 10 ocular divisions?	Stage micrometer (scale B) is now replaced by a slide. From your calculations in C, what is the length of the chain of streptococci?

Figure 2-1 Standardization of the ocular micrometer and its use in measuring bacterial cells. 1 micrometer (μm) = 1 $\times$ 10^{-6} meters.

calculate the number of micrometers in each division of the ocular scale. Record your calculations.

4. Replace the stage micrometer with the various stained cells provided and measure the dimensions of each type of cell with the ocular micrometer. Calculate your results in micrometers and record them on the report sheet.

QUESTIONS

1. If you did not have a stage micrometer, what might you use to calibrate your ocular micrometer? Explain.
2. What are the lengths in inches of the bacterial cells you measured?

Microscopic Measurement

Name _____

Desk No. _____

From your observations, complete the following:

_____21_____ ocular divisions = _____20_____ stage divisions

_____9.95_____ ocular divisions = 1 stage division = ___10___ μm

1 ocular division = ___10.52___ μm

What were the dimensions of the microbial cells observed?

Yeast cell _____ μm

Bacterial cell _____ μm

Protozoon *(Paramecium)* _____ μm

Alga *(Euglena)* _____ μm

amoeba _210.4 um_

staphylococcs aureus _10.52_ um

3. *Microbial Flora of the Oral Cavity*

As many as 300 species of bacteria may have been isolated from the human oral cavity (Figure 3-1). Two protozoa, *Entamoeba gingivalis* and *Trichomonas tenax*, and a yeast, *Candida albicans*, are also common. Streptococci, spheres in short chains, have been isolated from all sites in the mouth and comprise a large proportion of the normal oral flora. The species *S. mitior* is most common in dental plaque, in which rods and filaments are also common. Spirochetes (spiral-shaped bacteria) are common in scrapings from the gingival crevice and in dental plaque. The flora in your mouth are interesting and personal subjects with which to practice microscopy and are particularly good subjects to compare the image qualities of transmitted light and phase-contrast microscopy.

PROCEDURE

1. Take a toothpick and obtain a scraping from between two of your teeth or between your teeth and gums.
2. Place a drop of saline solution on a clean microscope slide and mix it with your toothpick scrapings so that the drop becomes slightly milky.
3. Very carefully place the edge of a clean coverglass at the edge of the drop and lower it slowly over the drop so that no bubbles form.
4. Place the slide on the stage of your microscope and adjust it for transmitted light microscopy, following the appropriate directions for your microscope.
5. Examine the specimen carefully and see if you can identify three or more morphological types of bacteria. The spirochetes may be too thin and fast for you to see clearly. You may also see the flagellate *Trichomonas tenax* and some of your own epithelial and red blood cells. On the report sheet, illustrate and label what you have seen.
6. As an optional step to be done in conjunction with Exercise 4 (or instructor's demonstration), after you have seen all the cell types you can identify, switch to phase contrast. Follow the directions appropriate to your microscope. Identify the same cell types again.

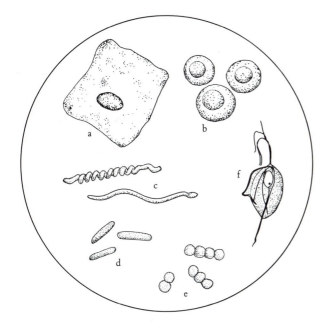

Figure 3-1 Some common morphological types of cells found in the human oral cavity: epithelial cells (*A*); red blood cells (*B*); spirochete (*C*); *Bacillus* (*D*); *Streptococcus* (*E*); and *Trichomonas tenax*, a protozoan (*F*).

QUESTIONS

1. Given the choice, which form of light microscopy, transmission light or phase contrast, would you use to study living cells? Why?
2. Why is the mouth such a good habitat for microorganisms? What are the sources of nutrition for the organisms found there?
3. Would you expect to find the same microbes in saliva and in plaque?

Microbial Flora of the Oral Cavity

Name _____

Desk No. _____

Illustrate and label what you have seen at 1000× in your mouth preparation.

Observation by ordinary (transmission) light microscopy

Observation in phase microscopy

4. *Phase Microscopy*

As you found out for yourself in the previous exercise, living bacteria and unstained cells are difficult to detect because there is little contrast between them and their aqueous environment. A number of specialized light microscopes are commercially available that provide contrast without staining: the phase-contrast, interference, and dark-field microscopes. Of these, the phase-contrast microscope is the most common instrument for studying unstained living cells. Interference microscopes are more costly and difficult to use, and dark-field microscopes, once popular, have largely been replaced by phase-contrast and scanning electron microscopes.

THE PHASE MICROSCOPE

Small differences in the *thickness* and *chemical nature* (index of refraction) of various parts of a specimen (e.g., nucleus and cytoplasm) retard part of the light (shift light in phase) as it passes through the specimen. Our eyes cannot detect phase shifts, but the phase microscope contains optical components that change phase differences into differences in brightness. To effect this change, an *annular phase ring* is placed in the optical system below or between elements of the condenser, and a matching **phase-shifting element** is placed at the rear focal plane of each objective (Figure 4-1). There are two

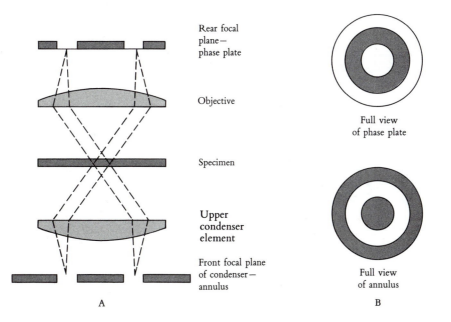

Rear focal plane — phase plate

Objective

Full view of phase plate

Specimen

Upper condenser element

Front focal plane of condenser — annulus

Full view of annulus

A

B

Figure 4-1 Phase-contrast microscope (dark contrast). *A*: Longitudinal section. *B*: Horizontal section.

methods of phase microscopy: bright contrast, in which the object appears as a bright image against a darkened background, and dark contrast, in which the object appears dark against lighter surroundings (Figure 4-2). The location and combination of light-retarding and light-absorbing material on the diffraction plate determine the type and degree of contrast.

THE INTERFERENCE MICROSCOPE

Various manufacturers use different systems to obtain interference, but the effects are similar. An interference microscope converts small differences in optical path length, indices of refraction, and qualities that cause rotation of light in different parts of the specimen into differences in *brightness* and *color*. All the light that enters the interference microscope is **polarized** by a filter located under the condenser (Figure 4-3). A prism or crystal plate, also located below the condenser, splits the light into two perpendicularly polarized components before it goes through the specimen. Above the objective and below the ocular is a third special element, a prism or crystal plate that recombines the split beams.

THE DARK-FIELD MICROSCOPE

A special type of condenser converts a bright-field microscope into a dark-field microscope. The light from the condenser is directed toward the specimen in a wide-angled hollow cone (Figure 4-4). Only light that hits the specimen and is reflected up into the objective forms the image. A specimen in dark field appears bright and self-luminous against a black background, much as does the moon against the dark sky. Dark-field microscopes are particularly useful for examining living unstained microorganisms suspended in fluid preparations.

USING THE PHASE MICROSCOPE

Most phase microscopes are of advanced-student- or research-grade design. The condenser must be properly focused and centered *before* beginning the next step. If in doubt, return to the directions for adjusting the microscope for transmitted light field (page 7).

1. Open the substage diaphragm all the way.
2. Look at the objectives. Note that each has a different colored ring around it (or a num-

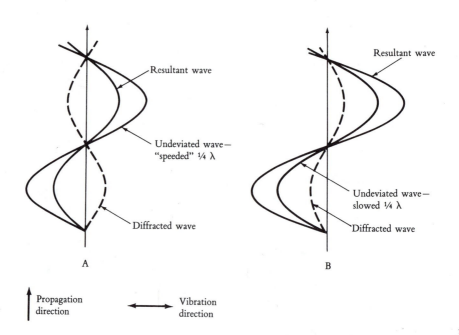

Figure 4-2   Comparison between dark phase contrast (*A*) and bright phase contrast (*B*).

Ocular

Analyzer

Movable modified
Wollaston prism

Back focal plane
of objective

Objective

Specimen

Condenser

Front focal
plane of condenser

Modified
Wollaston prism

Polarizer

Field
diaphragm

Light source

Objective

Object

Paraboloid
condenser

Objective

Object

Cardioid
condenser

Figure 4-3 Schematic representation of the light path in one design of an interference microscope.

Figure 4-4 Schematic representation of the light path in two designs of dark-field condensers.

B

A

Figure 4-5 Aperture viewing prism (*A*) or telescope (*B*) used in aligning the phase rings in the condenser with those in the objective. (Courtesy of American Optical Co.)

bered inscription on it). Rotate the knurled substage turret ring. Note that the ring has stops marked with the same colors (or numbers) as the objectives. The no-color stop or 0 is used for bright-field or "ordinary" light microscopy. If you cannot rotate the turret ring, pull out the phase-ring centering screws, which may have been pushed in accidentally (some designs have springs to prevent this from happening).

3. Rotate the correct phase ring for the objective into place.
4. Swing in the aperture viewing prism, or replace the ocular with a telescope, and focus on the phase ring (Figure 4-5).
5. If the ring in the turret is not centered, center it by means of the phase-ring centering screws, which you pulled out in Step 2 (Figure 4-6). *Note:* Zeiss microscopes have sliding cams instead of screws.

6. Disengage and pull out the phase-ring centering screws and flip out the aperture viewing prism or remove the telescope.
7. If you change objectives, repeat Steps 2–6. For each power, the rings are generally concentric, but centering must be checked because the rings are more critical as you go to higher powers.
   Note: Because of design and construction mistakes by some manufacturers, it may be necessary to raise the condenser to fill the phase rings properly at 100X. This problem is found in many, but not all, microscopes. These mistakes will put weird color fringes and halos on your image. Flip in the aperture viewing prism and raise the condenser until the rings of the objectives and the condenser are filled. If you have doubts, ask your instructor.

Figure 4-6 View of the substage of a phase microscope showing the location of phase alignment screws. (Courtesy of American Optical Co.)

PROCEDURE

1. Make a slide of the microbial flora of your oral cavity using the directions in Exercise 3.
2. Set up the phase microscope and observe the slide.

5. *Fluorescence Microscopy*

Fluorescence microscopy is growing in importance in biology because it is easy to use and because a remarkable degree of specificity can be visualized by linking a fluorescent dye (fluorochrome) to an antibody that is raised against a particular macromolecule of interest. Fluorescence is a phenomenon that occurs in substances that can absorb a quantum of light at one wavelength and, after a brief delay (less than 1×10^{-6} second), reemit a quantum of light of a longer wavelength. A number of biological molecules **autofluoresce** in the fluorescence microscope; for example, the chlorophyll in algal chloroplasts autofluoresce red when excited by the appropriate filters in a fluorescence microscope.

Fluorescein isothiocyanate (FITC), the most commonly used fluorochrome, absorbs deep blue light and emits green light. Rhodamine compounds, another series of common fluorochromes, absorb green light and emit red light. Fluorescent microscopes have lamps (mercury or tungsten–halogen) that emit strongly at the short end of the visible spectrum. The objectives in a fluorescence microscope have the highest possible numerical aperture so that they can collect the faintest fluorescence, and they are built of glasses and cement that do not themselves fluoresce. A fluorescence microscope has two special interference filters: an exciter filter and a barrier filter. Particular exciter-filter–barrier-filter combinations are used for particular fluorochromes. The exciter filter, placed between the light source and the specimen, allows only light of shorter wavelengths to pass through to the specimen (Figure 5-1). The barrier filter, placed in the optical system between the objective and the oculars, blocks the passage of light of short wavelengths and allows the passage of light with longer wavelengths. Effectively, the barrier filter blocks all the light that passes through the excitation filter; thus, if there is nothing in the object plane that fluoresces, the field is totally dark. But objects that fluoresce stand out like stars in a moonless sky.

Two designs for fluorescence microscopy are common: (1) *transmitted-light excitation* and (2) *epifluorescence* (also known as *incident-light excitation* or *reflected light excitation*) (Figure 5-2). The design of transmitted-light fluorescence microscopes is very similar to research-grade transmission microscopes; they differ only in light source and the mechanical means for holding exciter and barrier filters. Some of the early transmission fluorescence microscopes used dark-field condensers. Because of the fading of fluorescence, an irreversible photochemical effect caused by the exciting radiation, a light shutter is often added to the system to limit exposure of the specimen to UV light. In epifluorescence microscopes, the light source is mounted above the stage (Figure 5-2B). They do not use a condenser, but instead, the microscope objective is used both to focus the short-wavelength light on the specimen and to gather the fluorescent light emitted from it. A chromatic beam splitter reflects light onto the specimen but is transparent to the light fluoresced from the specimen. Epifluorescence tends to be brighter and easier to use than transmission fluorescence microscopy. It is often possible to leave the room lights on while using an epifluorescence microscope.

PROCEDURE

1. Take a drop of culture of *Euglena gracilis* or *Chlorella* sp. and place it on a microscope

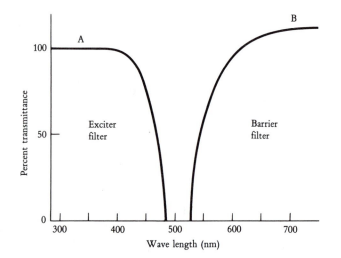

Figure 5-1   Schematic representation of light transmitted through a fluorescence exciter filter (*A*) and a barrier filter (*B*).

Figure 5-2 Schematic representations of two common fluorescence microscope designs. *A*: Transmitted-light excitation. *B*: Incident light excitation.

slide. (The cell density should be about 1×10^3 cells per milliliter. If it is higher, dilute it first.) Place a coverglass carefully over the drop so that no air bubbles are entrapped.

(a) As an alternative, take a drop of 1% acridine orange and add it to a culture of any bacteria that has a density of approximately 1×10^4 cells.
(b) Incubate for 30 minutes.
(c) Centrifuge cells and decant supernatant.
(d) Replace with fresh medium and mix thoroughly.
(e) Centrifuge cells and decant supernatant.
(f) Resuspend in fresh medium.
(g) With this suspension, repeat Step 1.

2. Place the slide on the stage of an epifluorescence microscope and lower a 10× objective close to the cover glass.
3. Raise the objective and focus on the algae.
4. Switch to the 40× objective and focus again.
5. Move the slide so that new specimens are in the field and with a stopwatch time the occurrence of fading.
6. On the report sheet, illustrate what you saw.

QUESTIONS

1. How could you use a fluorescence microscope to see if an algal culture was contaminated with bacteria?

2. Why is epifluorescence microscopy generally brighter than transmission fluorescence microscopy?

3. Why are exciter filters and barrier filters used in particular combinations?

Fluorescence Microscopy

Name _____

Desk No. _____

Illustrate what you saw in the fluorescence microscope.

6. *Demonstrations of Electron Microscopes*

There are two basic types of electron microscopes: the transmission electron microscope (TEM) and the scanning electron microscope (SEM). Both use electrons instead of light to form images. Because electrons have a very short wavelength (~5 picometers), the resolving power of the electron microscope is more than 100 times that of the light microscope, and it is possible to magnify images more than 200,000 times. The electron beams are focused by electromagnets instead of glass lenses. Living cells cannot be studied in either type of electron microscope because the air in the column in which the specimens are placed must be evacuated to prevent the electron source (filament) from rapidly burning out and to keep the electron beam from interacting with air molecules. Specimens are usually stained or coated with dense metals and images are viewed on a screen or a photographic plate.

THE TRANSMISSION ELECTRON MICROSCOPE

In its general design, the transmission electron microscope is similar to an upside-down light microscope (Figure 6-1A and B). Elaborate preparation of specimens is usually necessary for the TEM. Cells are chemically fixed with glutaraldehyde, dehydrated, and then embedded in plastic (Figure 6-2). Extremely thin slices of the plastic are cut with diamond or glass knives and then placed on supporting copper screens called grids. *Freeze etching* the specimen enables you to examine the internal cell components without the artifacts that can be produced by chemical fixation. In this technique, the specimen is rapidly frozen at $-100°C$ and then a glass knife is used to cut the surface; it is then freeze-dried and coated with carbon before being examined. Special techniques have been developed for making it possible to locate chemical constituents of cells in TEM. For example, antibodies labeled with electron-dense ferritin (iron-containing) or gold can be combined with particular cell antigens so that they can be seen in the transmission electron microscope.

THE SCANNING ELECTRON MICROSCOPE

The scanning electron microscope (SEM) is used for detailed examination of the external surfaces of microorganisms. The specimen is placed on a solid flat disc (stub) and then sputter-coated with gold or platinum under a vacuum. The stub is then placed in an electron microscope in which the electrons are focused into a needle-sharp probe only 5–10 nm in diameter (Figure 6-1C). The probe scans the specimen in a manner similar to a television camera. Because of the fineness of the beam and the short wavelength of electrons, the probe can enter very narrow openings and resolve details as small as 10–20 nm with good magnifications of more than 10,000×. Although the SEM cannot resolve details as small as a TEM can, or magnify a specimen as much, it has a depth of field almost 500× greater. The image on an SEM screen is formed by secondary electrons backscattered from the specimen and magnetically deflected to a detector. The secondary electron detector is synchronized with the probe beam and the image seen on the high-resolution television monitor represents successive areas that have been traversed by the probe beam. The scanning time of the scope can be adjusted to allow time for adequate concentrations of electrons to develop an image of the specimen at each point.

It is very easy to prepare materials for the SEM compared with the procedures needed to prepare materials for the TEM (Figure 6-3). The following exercises can be done by individual students or by instructors for demonstration at institutions with appropriate facilities.

PROCEDURE

Fluids

1. Put on protective gloves and bring a culture or natural water sample to a fume hood.
2. Add 2–6% (V/V) glutaraldehyde in 0.02–0.1 *M* buffer at pH 7. Allow to fix for 30 minutes.

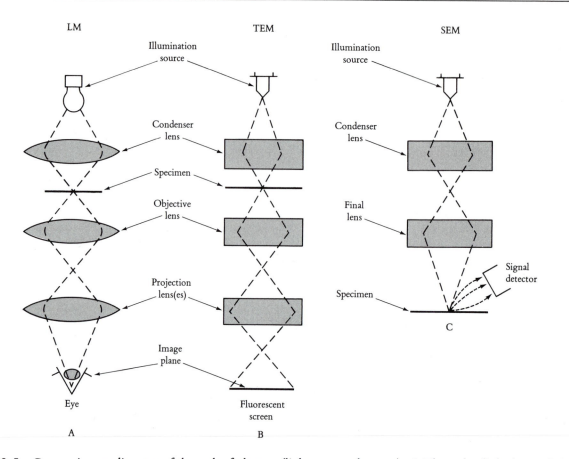

LM   TEM   SEM

Figure 6-1   Comparative ray diagrams of the path of photons (light rays or electrons). *A:* Through a light (transmission) microscope. *B:* Through a transmission electron microscope. *C:* Through a scanning electron microscope.

3. Filter sample through Nucleopore filter. (This type of filter is preferred because it has a flat surface and microorganisms stand out against it in the SEM.)
4. Flush filter with buffer solution.
5. Take an appropriate stub for the SEM you will be using and place a square of double-coated tape on it.
6. With a forceps, remove the Nucleopore filter and place it on the double-coated tape. Using the forceps, press the filter on the tape to make sure there is good adhesion.
7. Sputter-coat the specimen with a 10–20 nm layer of Au or Pd.
8. Give your specimen to the SEM operator to place in the column. Follow the directions the operator gives for examination of your specimen.

Note: Most prokaryotic microorganisms and many eukaryotic microorganisms with rigid cell walls and tests do not need to go through a step called **critical point drying**. This process prevents the shrinking and shriveling that occurs when organisms with thin cell walls or thecae are prepared for the SEM.

Solids

1. Put on protective gloves and work outdoors or with a fume hood.
2. With forceps take a leaf from an underwater plant, a stone, or bit of debris from an aquatic habitat and place it in a disposable plastic beaker with 2–6% (V/V) glutaraldehyde in

Tissue dissected out and placed in fixing solution.

After washing, the tissue is dehydrated by placing it in higher and higher concentrations of acetone or alcohol.

Tissue is now placed in dilute solution of plastic imbedding media.

Specimen vial

Specimen

Specimen holder for microtome

When the plastic is hard, the block is trimmed and is ready for sectioning.

Tissue is placed in final imbedding mixture and the plastic is polymerized in an oven.

Sections are cut on an ultramicrotome with a glass or diamond knife the sections are floated off the edge of the knife onto the surface of a water trough.

The sections are picked off the surface with a copper grid.

After the sections dry, they are ready for viewing in the electron microscope.

Figure 6-2 Preparation of microbial cells for examination in a transmission electron microscope (TEM).

Centrifuge and
rinse in buffer

Microorganisms fixed
in glutaraldehyde 2% in
ice bath in a fume hood

Post fix in 1% OsO$_4$ in
an ice bath in a fume hood

Rinse and
dehydrate
in a graded
series of
ethyl alcohol

Amyl
acetate   100%   95%   80%   70%   50%

Centrifuge and decant at each step

Critical point dryer

Mount
on stub

Sputter coater

Examine
in SEM

Figure 6-3 Preparation of materials for examination in the scanning electron microscope (SEM).

0.02–0.1 *M* buffer at pH 7. Allow to fix for at least 30 minutes (can be left for hours or days if refrigerated).

3. Remove specimen with forceps, rinse in buffer, and let air-dry.

Note: Some types, but not all microorganisms, may shrink and shrivel at this step. Better preparations must be critical-point-dried.

4. Take an appropriate stub for the SEM you will be using and place a square of double-coated tape on it.
5. With a forceps, place your specimen on the double-coated tape. Press hard so that there is good adhesion.

6. Sputter-coat the specimen with a 10–20 nm layer of Au or Pd.
7. Give your specimen to the SEM operator and follow the instructions given for examination of your specimen.

QUESTIONS

1. Why is it necessary to sputter-coat specimens before placing them in the SEM?
2. Why is it unnecessary to critical-point-dry most bacteria before observing them in the SEM?

T W O

Culturing Microorganisms

In nature, microorganisms exist as mixed populations of many different types. However, our knowledge of microbiology has come through study of isolated species, grown in environments free from contamination by other living forms. Such techniques are not required for studying higher plants and animals; they are unique to the study of the microbial world.

Like all other living forms, microorganisms need suitable nutrients and favorable environments. The **culture medium** must contain essential nutrients for the growth of a microbial culture, and it must provide suitable surroundings for growth: the proper pH, osmotic pressure, atmospheric oxygen, and other factors. If you consider the many materials that microorganisms can spoil, you will realize that many substances can serve as culture media.

Since most of our laboratory studies will be made with pure cultures, that is, with single species of microorganisms, we must be able to sterilize culture media and maintain them in sterile condition (free of living forms), and we must be able to inoculate a sterile medium with a pure (axenic) culture of microorganisms without contaminating it. This last procedure is referred to as **aseptic technique**.

Since newly prepared medium contains a mixture of microorganisms from its ingredients and the surfaces of utensils and glassware, it must be sterilized, that is, treated by heating or other methods until all these cells are destroyed. Prior to sterilization, a tube containing medium is usually loosely capped with one or a variety of fitted metal or plastic closures (Figure II-1). This prevents the entry of more contaminants but permits free interchange of air or other gases.

To start a bacterial culture, a number of cells (the **inoculum**) are transferred (inoculated) into a sterilized medium.

In the inoculation procedure the needle or loop that is used to transfer microorganisms should be heated to redness by **flaming** immediately *before* and *after* the transfer. Flaming destroys living forms on the surface of the needle or loop. Hold the needle *down* in the flame to heat the whole needle and the lower part of the handle (see Figure II-2).

During transfer, hold the tube in the left hand and the plug or cap between the fingers of the right hand. (*Caution:* Never lay a plug or cap down.) Hold the tube as nearly horizontal as feasible during transfer and do not leave it open longer than necessary (see Figure II-3). The mouths of the tubes from which cultures are taken and into which they are transferred should also be passed through the burner flame immediately *before* and *after* the needle is introduced and removed. In addition to destroying organisms on the lips of the tubes, flaming creates outward convection currents, which decrease the chance of contamination. Since most of the microbiologists' work is with pure cultures, you should master techniques of inoculation early in the course.

After inoculation, a bacterial culture is stored or incubated in a suitable environment for growth. *Growth*, in this case, means the development of a population of cells from one or a few cells. The mass of daughter cells becomes visible to the naked eye either as cloudiness (**turbidity**) in liquid broth or as an isolation population (a **colony**) on solid media. The appearance of a growth is one way of identifying microbial species.

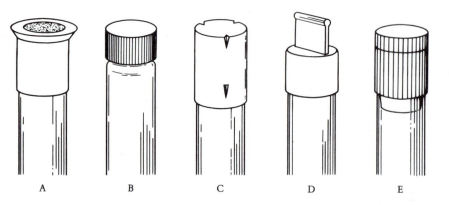

Figure II-1   Common culture tube closures. *A*: Sherwood Bacti Capalls®. *B*: Screw cap. *C*: Morton closure. *D*: Rubber culture tube caps. *E*: Sherwood polyethelene culture tube cap.

Figure II-2   Flame the inoculating loop before and after inoculation.

Figure II-3   Correct method for holding tubes and plug when transferring cultures.

An easy way to make a solid medium is to add a solidifying agent to a broth medium, which then hardens as it cools. The most common solidifying agent is **agar**, a substance obtained from marine algae and available commercially in dried purified form. Agar is easily dehydrated, can be mixed into media prior to heat sterilization, and the solid can be stored in test tubes or flasks. It is so easily handled in the laboratory that it seems almost tailored for the bacteriologist's special use. Although various agars differ considerably in their physical properties, their melting point is 97 to 100°C. Thus, solid agar media can be liquefied by boiling. On cooling, media containing agar solidify at about 42°C. If they are to be inoculated before hardening, they are cooled to 45 to 47°C, a temperature that is not harmful to most bacteria for short times. After solidifying, agar media can be incubated over the entire range of temperatures a bacteriologist is likely to use (up to 70°C) without melting.

Chemically, agar is a galactan, a complex carbohydrate composed of galactose molecules, which is not subject to breakdown by most bacteria. It is usually used at a concentration of 1.5%. A concentration of 1.8% gives a harder, less easily gouged medium. Except for some marine bacteria, agar is not a nutrient but only a hardening agent. Other hardening agents for special purposes are **gelatin**—which is used at a concentration of 12 to 15%, is liquid above 25°C, and is **hydrolyzable** (liquefiable) by many bacteria—and **silica gel**—a substance used in autotroph cultivation, for which organic matter must be excluded from the medium.

The following series of exercises will acquaint you with nutrient broth and nutrient agar, the most common forms of culture media, and their use in culturing bacteria. You will also learn the essentials of transferring microbial cultures aseptically and methods of isolating pure cultures.

7. *Broth Culture*

A simple way to obtain bacteria is to grow them in a test tube in a broth medium. There are numerous recipes for broths, and the type you use depends on the bacteria you want to grow. All liquid media, however, must provide the proper physical and chemical environment and nutrient substances in water solution.

Bacterial growth in broth becomes visible in several ways:

1. **Turbidity.** This condition is a cloudiness, more or less dense.
2. **Pellicle formation.** A small mass of cells floats on top of the broth.
3. **Sediment.** A deposit of cells rests at the bottom of the broth culture but swirls up if the tube is tapped gently (Figure 7-1).
4. **Slime.** If the cells do not separate as they rise from the bottom, slime may be present.

If dissolved gas is present, gas bubbles evolve if the tube is mixed or a hot loop is inserted.

PROCEDURE

Treat four tubes of nutrient broth.

1. Label one tube "Control" and leave it uninocculated. Do not remove the cap.
2. Remove the cap from one tube and add a small amount of dirt or other sources of microorganisms. Replace the cap.
3. After flaming a wire loop and the third tube of broth, inoculate with the pure culture of *Escherichia coli*. Since large numbers of microbes are easily transferred, there is no need to shake the loop. Repeating the procedure, inoculate the fourth tube with *Micrococcus luteus*.
4. Incubate the four tubes at 30°C until the next laboratory period.
5. Illustrate and describe the growth on the report sheet.

Observation

Examine the broth cultures for evidence of growth. Do not shake the tubes before making your initial observations for pellicle formation or sediment. Most bacteria reach the end of log growth phase by 24 – 48 hours in nutrient broth.

Note: Some heavy cells such as yeasts settle out completely, leaving the upper part of the broth sterile or nearly so. As a routine precaution during transfers, be sure the inoculum is in suspension.

As soon as you learn the procedure, you must develop speed in transferring cultures because delay in replugging tubes greatly increases the chances of contamination.

Figure 7-1   Method of tapping tube to suspend sediment.

QUESTIONS

1. When would it be necessary to use aseptic techniques when inoculating a tube of nutrient broth with dirt?

2. Why do some microorganisms characteristically form a growth of the pellicle type? What environmental factors might alter the formation of a pellicle?

REPORT 7

Broth Culture

Name _____

Desk No. _____

Illustrate and briefly describe the growth in each of the broth cultures.

Control

Inoculated with dirt

turbidity sediment

Escherichia coli

sediment

Micrococcus luteus

sediment

8. *Agar Slope*

An **agar slope** (also called a **slant**) is a test tube of agar medium that is placed at an angle to cool. The contents of a tube treated this way harden with a sloping surface that is easily inoculated with a loop or a needle. The agar slope provides a convenient surface for culturing microorganisms, especially aerobic and facultative anaerobic forms. Cultural characteristics such as pigment formation are most easily observed on slope cultures.

The agar slope and the **agar stab**, another method of culturing microorganisms on solid media, are common ways of maintaining **stock cultures** (see the discussion titled "Preservation of Laboratory Cultures" at the end of this section of the manual). The agar stab is advantageous when facultative anaerobic culture conditions are needed.

Inoculating needle

Agar slope (or slant) for inoculating in tube

Figure 8-1   Inoculation of the agar slopes.

PROCEDURE

1. Melt three tubes of nutrient agar in boiling water. Cool them in an inclined position.
2. When the medium is cold and solid, inoculate one surface with *Escherichia coli*, using a needle (Figure 8-1). Move the needle gently

back and forth on the agar surface from the butt (lower, or bottom) to the top, taking care not to gouge the agar. Inoculate the surface of a second tube with *Micrococcus luteus*. Leave a third tube uninoculated as your control tube.
3. Incubate all three culture tubes at 30°C until the next laboratory period.

Observation

Examine the mass of surface growth that has developed (Figure 8-2). Note that the growth of *E. coli* is very different on solid medium from that in liquid medium, which you observed in Exercise 7. Illustrate your observations on the report sheet.

QUESTIONS

1. Why is it important to keep the inoculating needle from gouging the agar surface?
2. Why don't bacteria (especially motile forms) grow throughout an agar medium, which is 97% H_2O?
3. When using slope cultures to study cultural characteristics, a needle is preferable to a loop for inoculating. Why?

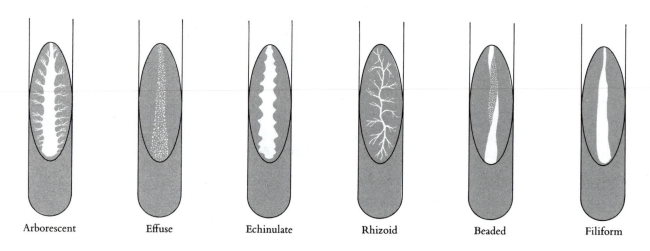

Arborescent   Effuse   Echinulate   Rhizoid   Beaded   Filiform

Figure 8-2   Different types of growth on agar slopes.

Agar Slope

Illustrate and label the growth observed.

Control

Mirococcus luteus

Escherichia coli

Effuse

Effuse

9. *Pure Culture Techniques*

Adding a solidifying substance to a broth medium that contains bacterial cells traps each cell in place. Instead of floating around as cells do when they multiply in a liquid medium, they produce a visible mass of cells, a fixed **colony**. If the cells are trapped some distance apart, each viable cell or clump of cells develops into a separate colony. This fact is of immense importance and underlies certain techniques for using solid media to identify and purify cultures, advantages not offered by liquid media.

Consider, for example, the problem of trying to separate mixtures of various types of bacteria in a liquid medium. The cells are too small to be picked out individually except by tedious, uncertain dilution techniques or by highly complicated procedures of single-cell isolation. However, if the cells are separated by dilution and then anchored in a solid medium in a petri plate, they form colonies that can be isolated in separate test tubes of nutrient medium. Furthermore, if the dilutions are carefully measured, the colonies that develop can be counted and the number of bacteria in the original sample calculated. (You will do this later.)

Note: Colonies of various microorganisms differ in size, shape, texture, and color; therefore, colony appearance is a valuable clue to the identity of a culture and the confirmation of its purity.

In this exercise you will isolate pure bacterial cultures by using the **streak-plate** and **pour-plate methods**.

STREAK PLATE

You can apply a microbial culture to the surface of agar in a petri plate and spread them with a loop or a bent needle. This is called **streaking**, and a plate so prepared is a **streak plate** (Figure 9-1).

There are several techniques for streaking plates. Two are described and illustrated here, and both produce excellent results if properly done (Figure 9-2). The objective of streaking is to produce well-separated colonies of bacteria from a concentrated suspension of cells. The cells are closely packed at the start of a streak and will form colonies that run together; but as a streak continues, fewer and fewer cells remain on the needle, rub off on the agar farther and farther apart, and thus form increasingly separated colonies. A few hasty streaks do not produce separate colonies. A good plate results from many continuous movements of the loop or needle over the plate.

In both techniques, you begin with a drop of culture at one edge of the agar. In the first technique, you flame a streaking needle (bent like a

Figure 9-1 Streaking an agar plate.

Flame and cool the loop
between 1 and 2, 2 and 3,
3 and 4, 4 and 5.

Flame and cool the loop
between 1 and 2, 2 and 3.

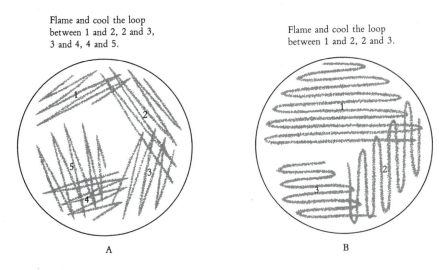

A   B

Figure 9-2   Motion of loop in two methods of streaking plates.

hockey stick) or a loop and cool it by jabbing it into the edge of the agar. Hold the petri plate cover in your left hand, partially covering the agar. With the drop on the edge away from you, streak the culture back and forth from edge to edge in parallel lines, moving toward you. When you have streaked one fifth of the plate, flame the needle, rotate the plate 45°, and continue streaking, following the pattern shown in Figure 9-2A. Flame the needle each time you change direction.

In the second technique, you start with the drop, streak halfway across the plate, flame the loop, turn the plate at right angles to the first streak, streak the plate halfway across the plate, flame the loop, rotate

the plate 90°, and streak halfway across half the plate. In this way you achieve the same result as before: dilution of the culture. Somewhere along the streaked paths, isolated colonies should appear Figures 9-3 and 9-4).

PROCEDURE

1. You are provided with three nutrient agar petri plates. With your inoculation loop, carefully streak a mixed culture on the three plates. Streak one plate by the first method (A) illustrated in Figure 9-2 and another

Figure 9-3   Isolated bacterial colonies on a streak.

Figure 9-4   A poorly streaked plate: no isolated colonies.

plate by the second method (B). Outline your streak pattern on the underside of each agar plate with a marking pencil to remind you of the orientation of your streaking.

2. Now that you have observed the principle of streak plating, design your own technique for streaking a culture and try it on the third agar plate.

3. Incubate the petri plates in an *inverted position* at 30°C until next period.*

   Note: Always label petri plates on the bottom with your name or initials, the date, and the identity of the contents.

4. After incubation, study the growth on your agar plates. Note the differences in colony size, shape, and appearance. Was your own streak technique successful in isolating individual colonies?

5. Illustrate the growth on the report sheet.

POUR PLATE

The technique of the pour plate gives you another way of obtaining pure cultures from a mixture of microbes. It differs from the streak plate in that the agar medium is inoculated while it is still liquid (but *cool*, 45°C), and so colonies develop throughout the medium, not just on the surface. This is an advantage for some purposes, for example, in studying the action of a streptococcal colony on red blood cells. A better distribution of colonies is also obtained in a well-made pour plate, and colonies are more easily isolated.

If cells in a mixture are "plated out," you face the problem of getting the proper concentrations of bacterial cells in the pour plates. Will the plates be so crowded with colonies that they are useless? Or will you have plates with no colonies at all? There is no accurate way of predicting the numbers of viable cells in a sample whether it is milk, a soil suspension, or a 24-hour bacterial culture. Therefore, always make several dilutions of the sample and pour several plates. You can expect that only

one or two of the plates will have useful numbers of colonies. The plates with too many or too few colonies can be discarded. However, experience will enable you to minimize wasteful dilutions.

The **loop-dilution procedure** (see Figure 9-5) usually yields useful plates from most samples. It is based on a roughly quantitative dilution of a sample in agar medium.

PROCEDURE

1. You are provided with three tubes of melted nutrient agar. Temper, or cool, this agar to 45–47°C.

2. Aseptically transfer 1 loopful of a mixed microbial culture to the first tube of melted agar. Mix the inoculum well by thumping the tube with your forefinger or by using a vortex mixer (Figure 9-6). The success of this method depends on thorough mixing, which evenly distributes the cells.

3. Transfer 2 loopfuls of the mixed inoculum and agar from the first tube to a second tube of melted agar and mix well.

4. Transfer 3 loopfuls from the second to the third tube of agar and mix thoroughly. (*Caution:* Since the drop in temperature from 45 to 42°C occurs rapidly, these steps must be carried out quickly before the agar solidifies.)

5. Pour each tube of agar into a separate petri plate, rotating this plate to distribute the agar evenly (Figure 9-7).†

6. Incubate the petri plates at 30°C until the next laboratory period.

7. Examine the colony growth on the plates.

8. Illustrate the growth on the report sheet.

---

*Because of the high percentage of water in agar, condensation can occur in petri plates during incubation. Moisture drips from the cover to the surface of the agar and spreads, resulting in a mass of confluent growth. To avoid this, petri plates are routinely incubated bottomside up.

---

†Several precautions are necessary to prevent contamination. First, when you take the melted agar from the water bath, wipe the outside of the tube with a cloth or paper towel. Otherwise, when you pour the agar plate, the water runs into the plate and introduces contaminants. Next, when you remove the tube cap to pour the agar, flame the mouth of the test tube to kill microorganisms on the outside lip. In pouring the agar from the tube to the plate, raise the cover of the plate on only one side, just enough to admit the mouth of the tube. You must also take care not to scrape the tube on the dish or its cover when pouring the agar.

After you have poured the agar into the plate, cover it immediately, pick up the plate, and gently tilt it from side to side to distribute the agar uniformly over the bottom. Replace the plate on the table and allow the agar to harden undisturbed until it is solidified and cold.

1 loopful   2 loopfuls   3 loopfuls

Mixed broth culture

Pour into petri dishes
after mixing in tube

Tube #1
melted agar,
45°C

Tube #2
melted agar,
45°C

Tube #3
melted agar,
45°C

Dilution #1   Dilution #2   Dilution #3

Figure 9-5 The loop-dilution technique.

Figure 9-6 Using a vortex mixer.

Figure 9-7 Pouring an agar plate.

On the crowded plates the colonies tend to be very small and to run together so that growth appears as a cloudiness. Well-isolated colonies are larger, with distinct and characteristic form, shape, color, and texture.

Note also that colonies that happen to have grown on the surface of the plate are spread out, large, and probably circular. Colonies that have developed deep in the agar are relatively small and probably present a lenticular (◯) cross section, reflecting the physical limitation of growth in the confines of the agar.

It is obvious that in mixed populations of microorganisms from natural sources (intestinal contents, fermenting vegetable matter, soil, and so on), some species are present in much greater numbers than others. This points out a weakness in dilution techniques for isolation: Organisms present in small numbers can be diluted to the extinction point on plates crowded with more dominant species. The subordinate organisms can be recovered in some instances, for example, by using a repressing chemical that is specific to the dominant species. Its growth is prevented, which permits growth of the less prevalent species and isolations can be made (see Exercise 40).

QUESTIONS

1. Why do you cool melted agar to 45–47°C before pouring the agar plate? before inoculating?
2. What factors contribute to the differences in appearance of (a) isolated versus crowded colonies and (b) surface versus subsurface colonies?
3. What factors limit the size of a bacterial colony?

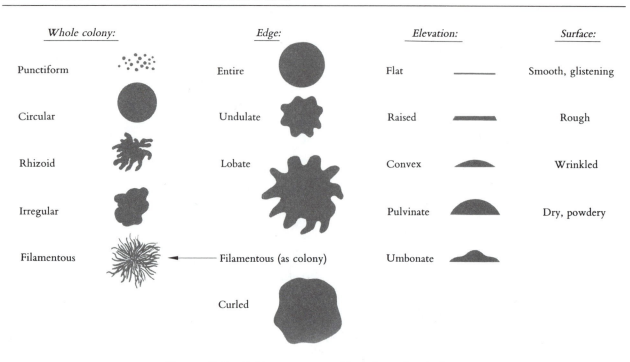

Figure 9-8 illustrates labeled colony characteristics:

Whole colony: Punctiform, Circular, Rhizoid, Irregular, Filamentous

Edge: Entire, Undulate, Lobate, Filamentous (as colony), Curled

Elevation: Flat, Raised, Convex, Pulvinate, Umbonate

Surface: Smooth, glistening; Rough; Wrinkled; Dry, powdery

Figure 9-8   Guide to terms used in colony descriptions.

ISOLATING A BACTERIAL CULTURE

Having successfully prepared streak and pour plates, you can now transfer cells from your various well-separated colonies to tubes of sterile broth. After incubation, cells from each plate should give you a pure culture, that is, a culture of a single species. The culture should be checked for purity by streaking it on a second plate. If all of the colonies that grow on the second plate appear the same, you probably have a pure culture.

PROCEDURE

1. From either the streak plate or the pour plate (or both), select three obviously different colonies (two surface and one subsurface). Mark them on the underside of the plate with a red pencil.

2. Flame an inoculating needle and allow it to cool for a moment. Lift the cover of the petri plate, touch the colony to be isolated with the needle, and transfer some of the colony to a tube of sterile broth. Observe the precautions of aseptic technique, flaming the lip of the tube and the needle before and after introducing the inoculum. Repeat the procedure for two other colonies.

3. Label each tube with an identifying number and then incubate the tubes at 30°C until the next laboratory period. You will use these cultures for staining exercises. Make observations of their growth on the report sheet. Figure 9-8 illustrates some colony types and their descriptive nomenclature.

4. During the next laboratory period, streak each of your broth cultures on a second nutrient agar plate. If all of the colonies on a plate appear to be the same, then you have probably isolated a pure culture. Figure 9-9 illustrates a rather obviously mixed culture.

Figure 9-9   Mixed culture.

PRESERVING LABORATORY CULTURES

A microbial culture can be preserved to avoid the necessity of subculturing or transferring it to fresh media at frequent intervals. To be suitable, a means of preservation should permit maximum survival of the population with little or no genetic and physiological change. In principle, all methods depends on bacteriostasis, that is, maintenance of the culture without growth or reproduction. Most methods employ refrigeration or desiccation (or both) to bring about dormancy.

Short-term preservation can be accomplished in a number of ways, depending on the characteristics of the particular species. Aerobes can be preserved for months on refrigerated agar slopes (see Exercise 8). Some facultative anaerobes, for which oxygen may be injurious, are preserved as refrigerated agar-stab cultures. The agar stab is prepared as illustrated in Figure II-4. A tube of sterile melted agar is solidified in an upright position; then the medium is inoculated by thrusting the inoculating needle into the center of the agar. During incubation, a core of growth develops. To exclude oxygen and to prevent evaporation, a seal of sterile mineral oil can be applied over the agar stab, after which the stab culture is refrigerated. Strict anaerobes are cultured and preserved in a strongly reducing medium such as thioglycollate broth (see Exercise 29). If acid is a metabolic product of the organism being preserved, the medium should be well buffered.

Refrigerated cultures of the sort just described remain viable for several months, but refrigeration is not suitable for long-term preservation because such stocks must be recultured periodically and, more importantly, because cultures that are refrigerated for long periods may mutate. Such difficulties can be largely overcome by other techniques. Cells that can survive freezing or dehydration can be maintained in a dormant state for long periods by being frozen or dried. Some species do not survive ordinary freezing or dehydration, but most bacterial cultures survive well if they are frozen in litmus milk or if they are mixed 1:20 with sterile glycerol and kept at −20°C or lower. Long-term storage can be accomplished by sealing vials of cells and holding them at −100 to −200°C in liquid nitrogen.

The most common method for long-term preservation of microbes is the freeze-drying technique called **lyophilization**. The microbial culture is suspended in a sterile milk, blood serum, or some equally protective substance in a glass vial or tube, quick-frozen with a mixture of dry ice and alcohol, and then thoroughly desiccated while frozen. Finally, the vial of dried culture is sealed (Figure II-5)

Desiccated cultures can be stored in the vials for several years. A high percentage of the microbial population survives lyophilization, and thus there is little or no selection for the more resistant genetic variants in the population. The high survival rate with this technique, compared with other methods of preservation, results from a combination of

Figure II-4 Preparing an agar-stab culture.

the protective effects of the suspending medium and of the desiccation of the culture in a frozen state.*

Figure II-5   Lyophilizing a culture.

To pump

---

*The American Type Culture Collection, 12301 Parklawn Drive, Rockville, Maryland 20852, maintains a vast number of microorganisms and viruses that are available at nominal cost to scientists.

Name _____

Desk No. _____

Illustrate the growth on the streak plate and the pour plate. Use arrows to show the direction of streaking. On the outside of each plate, draw each different colony type you find on your plate. Draw a line from your colony illustration to the place at which you found each colony type.

STREAK PLATE

punctiform
circular

POUR PLATE

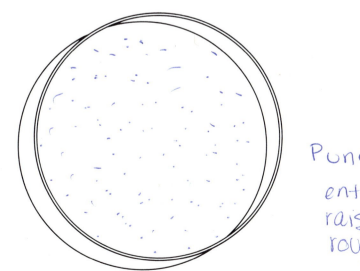

Punctiform

entire
raised
rough

Preparing and Sterilizing Media

MEDIA

The many types of media used in microbiology vary according to the needs of the particular organisms to be studied. A distinction between complex and defined media is sometimes made.

Complex media contain all necessary ingredients for growth of a microorganism, but they are in crude form, that is, not all the components of the media nor their exact quantities are known. Many components in complex media are acid or enzymatic digests of various plant tissues, meats, casein, and yeast cells that provide rich sources of polypeptides, amino acids, vitamins, or minerals. Examples of such components of media are peptones or tryptones, which are enzymatic hydrolysates of animal protein, or yeast extract, which is autolyzed yeast, rich in B-complex vitamins. Some carbohydrate is usually present in these crude digests, but many complex media are supplemented with sugar, usually glucose.

Defined media (holidic) provide the nutrients required for growth in the form of relatively pure chemicals of known concentration. They vary according to the nutritional needs of the microorganism. The essential components of a medium for growing nonfastidious heterotrophs, such as *Escherichia coli*, are inorganic salts and carbon and nitrogen sources. A suitable defined medium for *E. coli* might be composed of NH_4Cl, $MgSO_4$, KH_2PO_4, Na_2HPO_4, and glucose. Other essential elements such as iron, manganese, and copper are usually present, as contaminants of the chemicals, in sufficient amounts for growth. Glucose provides energy and a source of carbon, and the ammonium chloride is the nitrogen source. A more fastidious heterotroph, lacking the synthesizing capabilities of *E. coli*, might require the addition of various vitamins, amino acids, purines, and pyrimidines.

Culture media must be adjusted to a suitable pH. This is usually done before sterilization.

Culture media are usually sterilized by heating in an autoclave at 121°C under 15 pounds of steam pressure for 15 to 30 minutes.

THE MEANING OF pH

The number of hydrogen ions (H^+) in an aqueous solution depends on the extent of dissociation of the acids present. The pH is a logarithmic function, defined by the equation $pH = \log 1/(H^+) = -\log(H^+)$. Thus, a pH of 7 (pH 7) represents a hydrogen ion concentration (H^+) of 1×10^{-7}. This is the hydrogen ion concentration for pure water at 25°C and indicates neutrality on the pH scale.

A solution that is above pH 7 is more alkaline (has a lower hydrogen ion concentration) than water. A solution below pH 7 is more acidic than water. For example, a 0.000001 N solution of HCl has 10^{-6} moles of (H^+) per liter, or a pH of 6. Because the pH scale is logarithmic, one pH unit represents ten times as many hydrogen ions as the next higher unit. For example, 0.1 N HCl has 10^{-1} (H^+) or a pH of 1, whereas a 0.01 N HCl solution has 10^{-2} (H^+) or a pH of 2.

The pH of most culture media must be near neutrality for optimum bacterial growth. When first prepared, most media are slightly acidic and need pH adjustment. In the preparation of media, the pH can be determined either potentiometrically or colorimetrically.

Potentiometric methods of determining pH measure the potential difference between the solution tested and a standard. In most pH meters a glass electrode half-cell with a definite potential is used with a reference electrode half-cell having a liquid junction, usually saturated KCl. Potential difference is measured between the terminals of the two half-cells in volts and converted to pH on the meter scale.

Colorimetric methods of determining pH use various indicators in solution or on pH paper; the color of the solution or paper corresponds to a particular pH. Such indicators are compounds that exist in solution as proton donor–acceptor systems in which donor and acceptor differ in color. The full range of pH is covered by combining the various indicators, with different pK values (equal to the pH at which a buffer salt is 50% dissociated) as indicated in Table III-1.

The pH can be adjusted by titration of a portion of the medium. From the amount of acid or base needed for this titration, the amount necessary to bring the remainder of the batch to the proper pH can be calculated. For a medium whose pH changes during heating, adjustment must be made after sterilization; for this, sterile acid or alkali is added aseptically.

Table III-1   The pH Ranges of Indicators

Indicators	Full Acid Color	Full Alkaline Color	pH Range
Thymol blue (T.B.) (Acid range)	red	yellow	1.2–2.8
Bromphenol blue (B.P.B.)	yellow	blue	3.0–4.6
Bromcresol green (B.C.G.)	yellow	blue	3.8–5.4
Methyl red (M.R.)	red	yellow	4.4–6.0
Chlorphenol red (C.P.R.)	yellow	red	5.0–6.6
Bromcresol purple (B.C.P.)	yellow	purple	5.4–7.0
Bromthymol blue (B.T.B.)	yellow	blue	6.0–7.6
Phenol red (P.R.)	yellow	red	6.8–8.4
Metacresol purple (M.C.P.)	yellow	purple	7.4–9.0
Thymol blue (T.B.)	yellow	blue	8.2–9.8
Phenolphthalein	colorless	red	8.0–9.6
Cresolphthalein	colorless	red	8.2–9.8

10. *Preparing Culture Media*

The most common complex medium for culturing microorganisms is nutrient broth containing beef extract and peptone. Beef extract is prepared by boiling lean beef in water and then evaporating the stock to a paste. The major constituents are protein degradation products, organic bases, vitamins, and mineral salts.

Laboratory supply firms prepare a dehydrated broth of 0.3% beef extract and 0.5% peptone that can be reconstituted with water and supplemented with 1.5–2.0% agar if a solid medium is desired. Since such dehydrated products usually have a preadjusted pH of about 7, further adjustment should not be necessary.

In this exercise, you will prepare and sterilize nutrient agar. Using this agar, you will prepare plates and then make streak plates as a test of your mastery of the streaking technique.

PROCEDURE

1. Put 0.8 g of dehydrated nutrient broth powder into a 250-ml Erlenmeyer flask. Slowly add 100 ml of distilled water while swirling the flask gently to dissolve the powder. The solution should be clear.
2. When the nutrient broth powder is completely dissolved, weigh out and add 2 g of agar.

   Notice that the agar does not dissolve until the medium is heated to about 100°C. You can save a step in preparation by heating the medium to melt the agar and to sterilize at the same time.

3. The medium is then sterilized in the autoclave (see Exercise 11).
4. After sterilizing the nutrient agar, swirl it *gently* to disperse the agar evenly and cool it to about 45°C.
5. When the agar has cooled, aseptically pour four plates, approximately 15 ml per plate. If bubbles appear on the surface of the agar, remove the lid, and quickly flame the surface with an inverted Bunsen burner to remove the bubbles.
6. Incubate your labeled plates and flask for 48 hours at 30°C.
7. At the next laboratory period, observe your plates for contamination. Discard contaminated plates.
8. Streak your plates with the culture provided.
9. Put your plates, properly labeled, into the tray provided. Your instructor will incubate and then grade the plates for streaking technique.

QUESTIONS

1. Why was it unnecessary to adjust the pH of the nutrient broth? If *you* had prepared the beef extract from lean beef, would pH adjustment have been necessary? How would you adjust the pH of an agar medium?
2. Why has it been impossible to grow some types of microorganisms on a defined medium?
3. What are the advantages of complex media in the general culturing of microorganisms?

11. *Sterilization Methods*

Sterilization, the process of killing all living organisms in or on a material, can be achieved by exposing the material to lethal physical or chemical agents or, for liquids, by mechanically separating out the organisms. There are many practical methods for sterilizing materials, and the choice often depends on convenience and the nature of the materials to be sterilized.

HEAT STERILIZATION

Moist Heat

The most widely used equipment for sterilization is the **autoclave**, which can sterilize metals, dressings, glassware, and most, but not all, aqueous media. Autoclaves work like pressure cookers. Laboratory autoclaves are usually operated at a steam pressure of 15 lb/in² at a temperature of 120°C. An autoclave sterilizes most materials in 15–30 minutes, the variation in time being due to the surface-to-volume ratio of the material being sterilized.

Some Aspects of Pressure–Steam Sterilization

Temperature. Bacterial endospores are the forms of life most resistant to heat, and a killing temperature can be reached only if the steam is under pressure. A temperature of 121°C provides a good margin of safety if it is maintained for the proper length of time.

Moisture. Coagulation of bacterial protoplasm (proteins, enzymes, and so on) at moderate temperatures requires moisture, and as moisture is removed, the temperature required for coagulation increases rapidly. As steam becomes superheated, it also becomes drier, so the temperatures and exposure times required for sterilization increase to approach those of hot-air sterilization (170°C for one hour). Thus, excessively heated steam loses some of its efficiency as a killing agent. In addition, the increased temperatures can be detrimental to the materials to be sterilized.

Pressure. Pressure itself has no effect on sterilization in the ranges used with the autoclave. Pressure is useful only for heating steam to temperatures above 100°C.

Time. Time is needed for the steam to penetrate and heat materials to sterilizing temperatures. Even when sterilizing temperatures are attained, the spores (and vegetative cells) are not all killed at once. The death rate is constant at a given temperature, and for each unit of time of exposure to a lethal agent, a constant proportion of a given population is killed. Usually it takes 11–12 minutes at 121°C (moist heat) to kill the endospores of thermophilic bacteria. For some delicate foods or media it is useful to place the medium in a steam sterilizer for short periods, let it cool for several hours, and replace it in the steam sterilizer (intermittent sterilization). All the vegetative cells are killed in the first treatment. The spores present germinate between the first and second treatments.

Entrapped Air. The relatively cool air in the sterilizing chamber is more than twice as heavy as steam at sterilizing temperature. If not allowed to escape, a stratification results. Since air and steam are slow to mix, the difference in temperature between the upper and lower layers can be very great, hence the need for replacing all the air with steam. Even after mixing, the temperature can be below that required. The thermometer in the discharge outlet indicates that all the air is exhausted when it reads 100°C.

Nature of the Load. In general, bulky and more impervious materials require more time. It is preferable, therefore, to sterilize in the smallest convenient units, for example, in five 1-liter flasks rather than in one 5-liter flask.

Flasks should be plugged with cotton or capped with paper. If it is necessary to use rubber stoppers, screw caps, or plastic caps, they should be set in place loosely to allow air to escape, to prevent the containers from bursting or blowing off the caps as steam is generated, and to allow steam to penetrate easily.

Dry Heat

Dry heat is used to sterilize glassware, other heat-stable solid materials, and a few medium components or materials (e.g., rice starch and animal food pellets) that would become unusable if exposed to steam.

GAS STERILIZATION

The recent proliferation of plasticware as disposable syringes, petri dishes, culture tubes, plastic bags, plastic filtration devices, plastic-backed adhesive bandages, and many other items has hastened the development of a type of autoclave for heat-labile materials: the **gaseous autoclave** (Figure 11-1). Ethylene oxide is the gas most commonly used in such instruments. Economic decisions and the kind of plastic determine the sterilizing regimens in this predominantly commercial process. For example, if the humidity is kept between 20 and 50% and the concentration of ethylene oxide is kept to 720 mg/l, then sterilization can be achieved in 4 hours at 54.4°C (130°F) or in 8 hours at 37.7°C (100°F).

Ethylene oxide and other gases are increasingly important as sterilizing agents. Although many laboratories do not have the equipment necessary to demonstrate **gaseous sterilization**, you should be aware of its use and advantages. The use of ethylene oxide vapors under pressure, in special equipment resembling a modified autoclave, is becoming a common method of "cold" sterilization. Ethylene oxide is very toxic to viral particles, bacterial and fungal cells, and the most heat-resistant bacterial endospores. As a sterilizing agent, it is easy to handle with proper equipment and is relatively inexpensive. Unlike most toxic chemicals, it is relatively noncorrosive and nondeleterious to materials being sterilized. Residual amounts of it are easily removed by aeration.

Although ethylene gas itself is flammable, 10% ethylene oxide with 90% carbon dioxide or mixtures with freon are effective sterilizing agents that are nonflammable and nonexplosive. Its disadvantages include long exposure periods for sterilization (several hours), reactivity with components of media and certain types of plastics, and, as mentioned, the residue of ethylene oxide that remains after sterilization, which must be removed by aeration or by allowing the sterilized material to stand.

Figure 11-1 Ethylene oxide sterilizer. (Courtesy of American Sterilizer Co.)

Figure 11-2 Membrane-filter apparatus. *A*: Example of a reusable apparatus. *B*: Example of a disposable plastic apparatus.

RADIATION

A few commercial processes use radiation for cold sterilization of certain materials (for example, pharmaceuticals). **Irradiation,** is the use of high-energy ionizing radiation, which includes gamma rays from a cobalt-60 or cesium-139 source and cathode rays from electron generators and accelerators. Irradiation by ultraviolet light is not a satisfactory means of sterilization because of the low penetrating power of the wavelengths of the ultraviolet portion of the spectrum.

A

B

C

Figure 11-3 Swinney syringe filters. *A*: Reusable filter assembled and attached to syringe. *B*: Reusable filter component parts. *C*: Disposable filters.

FILTRATION

The principal method of sterilizing liquids that contain heat-sensitive components such as vitamins, serum proteins, and antibiotics is by filtration. Historically, microbiologists sterilized heat-labile media components by passing them through autoclave-sterilized filters made from diatomaceous earth, asbestos fibers, or sintered glass. Since these types of filters are hard to clean and have other disadvantages, they have been replaced by disposable cellulose acetate or polycarbonate membrane filters held in stainless steel, glass, plastic, and disposable plastic funnels (Figure 11-2). Suction can be used to draw the liquid through the filter. Another popular filtering device is the **Swinney filter** (Figure 11-3), which is used to sterile-filter small amounts of heat-labile liquids such as antibiotics and serum components as they are added aseptically to cultures or media. A pore size of 0.45 μm is conventionally used for sterilization. Marine microbiologists have found that many marine bacteria can pass through this pore size, so they use filters with a pore size of only 0.2 μm.

PROCEDURE

Autoclaves vary greatly from model to model and among manufacturers. Pay careful attention to the directions given you by your instructor.

Operation of the Pressure–Steam Sterilizer, or Autoclave

1. Load the sterilizer with the flasks of nutrient broth prepared in Exercise 10.
2. Close and lock the door (Figures 11-4 and 11-5).*
3. Open the operating valve (air outlet) to allow the air in the chamber to be displaced by steam.
4. Open the steam supply valve. This admits steam to the sterilizing chamber from the steam line.
5. Watch the thermometer until the temperature approaches 100°C (212°F) and then close the operating valve. Almost all the air

---

*Figure 11-5 shows a modern autoclave in which the entire sterilization procedure is regulated automatically.

Figure 11-4 The basic structure of an older model autoclave. (Courtesy of American Sterilizer Co.)

Figure 11-5 A modern autoclave. (Courtesy of American Sterilizer Co.)

and condensed water should have been discharged from the chamber by this time. (*Note:* On some models this valve closes automatically by thermostatic control.)

6. Continue to watch the thermometer because the temperature will keep rising. Pure steam at one atmosphere of pressure cannot exceed 100°C; it must be under pressure to go above 100°C. Most modern autoclaves, however, have adjustable pressure regulators by means of which the maximum pressure, and consequently the maximum temperature, can be controlled.

   Autoclaves for routine laboratory sterilization are usually set for 15 pounds pressure, giving a temperature of 121°C (250°F).

This is suitable for sterilization if maintained for 15 minutes. When the temperature reaches 121°C, begin timing sterilization.

Note that the thermometer measures the temperature of the steam in the discharge line. If air has not been completely removed from the autoclave, even though the pressure gauge indicates 15 pounds pressure, the temperature will not reach 121°C. Therefore, it is essential to measure sterilization time from the moment the thermometer reaches 121°C rather than from the time 15 pounds pressure is indicated.

7. To shut off the autoclave at the end of sterilization, all you have to do is to close the steam supply valve and wait until the pressure falls to zero on the gauge. If the materials that were sterilized did not contain liquid, but *only if they did not*, you can open the operating valve to release the pressure more rapidly. If liquids are in the chamber, sudden release of pressure causes them to boil up in the containers, wetting the plugs, and blowing them out. It is, therefore, necessary always to let the pressure fall gradually when liquids are being sterilized. Most modern autoclaves, many of which are entirely automatic, are equipped with automatic locking devices on the doors that do not permit the doors to be opened until the pressure has lessened.

8. After the pressure gauge indicates that there is no steam pressure, you may open the door and remove the load. If the materials can be damaged by prolonged heating or by evaporation, cool them promptly.

QUESTIONS

1. What is the theory upon which intermittent sterilization is based?
2. Is dry or moist heat more efficient as a sterilizing agent? Why?
3. What is the temperature in degrees F when it is 170°C?

List and discuss the special purposes of all the sterilization equipment you used and saw in demonstration in your laboratory or media preparation room.

Name the type of autoclave available to you for the sterilization of the media you made and list the instructions for its use. Underline any precautions that were given for your safety or to prevent ruining the medium.

F O U R

Staining Bacteria

SIMPLE STAINING TECHNIQUES

Bacterial morphology can be examined in two ways: by observing living, unstained organisms and by observing dead cells that are stained with dyes.

Living bacteria are almost colorless and lack sufficient contrast with the water in which they are suspended to be clearly visible. Staining the organisms increases their contrast with their surroundings so that they are more visible. Certain stains can help identify internal structures of cells that would otherwise be unseen. Further, use of the oil-immersion objective of the microscope to obtain the greatest magnification is more convenient with stained preparations than with wet mounts.

Although bacteria do not look greatly different from their surroundings, they differ *chemically*. Stain or dye reacts chemically with the bacterial cell but not with the background, enabling us to distinguish the bacteria. Thus, the main advantages of staining are that it provides contrast between microorganisms and their backgrounds,

(1) To differentiate between different forms of microorganisms

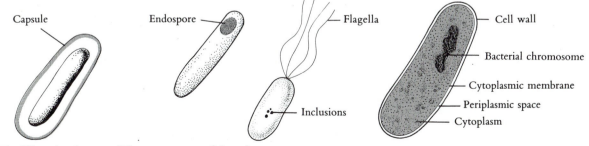

(2) To differentiate between different structures of the cell

Figure IV-1   Reasons for making and using stained preparations of microorganisms.

enables differentiation among various morphological types, and enables study of internal structures of the bacterial cell such as the cell wall, vacuoles, and spores.

A wide range of dyes is available to the bacteriologist today and is used in various modifications of basic staining techniques:

1. Simple stains
   (a) Basic dyes
   (b) Acidic dyes
   (c) Indifferent dyes

2. Differential stains
   (a) Gram stain
   (b) Acid-fast stain (Color Plate 1)

3. Structural stains
   (a) Endospore stain (Color Plate 2)
   (b) Capsule stain
   (c) Flagella stain (Color Plate 3)

12. *Direct Staining with Basic Dyes*

To understand how a dye stains a bacterial cell, you must first know what a dye is. Most dyes are salts, of which one of the ions is colored. A salt is a compound composed of a positively charged ion and a negatively charged ion. The simple dye methylene blue is actually the salt methylene blue chloride, which dissociates as follows:

$$MBC \rightarrow methylene\ blue^+ + chloride^-$$

The color of the stain is in the positively charged methylene blue ion.

Bacterial cells have a slight negative charge when the pH of their surroundings is near neutrality, which it generally is. The negatively charged bacterial cell combines with the positively charged methylene blue ion, with the result that the cell is stained. The difference in charge produces an affinity between the dye and the bacterial cell.

Dyes can be divided into two groups: basic and acidic. If the color is in the positive ion of the dye, we call it a **basic dye** or stain. Thus, methylene blue is a basic dye. If the color is in the negatively charged ion, we call it an **acidic dye** or stain.

In this exercise you will make stained preparations of the cultures that you isolated in Exercise 9 and also of saliva.

PROCEDURE

Preparing and Fixing Bacteria for Staining

Prior to staining, you must fix the material to be observed, that is, make it stick to the glass slide on which it is to be stained. If a preparation is not fixed, the film of cells washes off during the staining procedure. The method illustrated in this experiment is preliminary to most staining methods used for bacteria. In this exercise you will use heat to kill the cells and cause them to adhere to the slide.

1. With a wire loop place a small drop of each of the broth cultures isolated in Exercise 9 on separate clean slides.* If a culture is taken from solid media, place a small drop of water on the slide and thoroughly mix with it a *small* bit of the culture.
2. Spread the drop on the slide to form a *thin* film. Most students make the smear too heavy.
3. Collect some saliva in a sterile test tube. With a sterile inoculating loop transfer a loopful of it to a clean slide and spread to form a thin film.
4. Allow the slides to dry in the air or hold them high above a bunsen flame.
5. When the film is dry, pass the slide, film side up, three times through the bunsen flame. *Caution:* Too much heat distorts the shapes and structures of the microorganisms. The slide should feel warm but not hot against the back of your hand.

The purpose of fixation is to kill the microorganisms, coagulate the protoplasm of the cells, and cause them to adhere to the slide. The ideal fixing agent preserves the structures of the cells in their respective forms and positions without causing the appearance of artifacts (structures not present in the living cell). Although gentle heating is the most commonly used method of fixation, agents such as alcohol and various chemicals can also be used. (Figure 12-1 is an illustration of how to prepare a smear for staining.)

Staining with Basic Dyes

For staining you can use the basic dyes **methylene blue**, **crystal violet**, and **carbol fuchsin**. These dyes differ in their rate and degree of staining. Methylene blue reacts with the negatively charged cell at the slowest rate, taking 30–60 seconds to stain a microbial preparation properly. Crystal vio-

---

*It is imperative that a slide be clean. One suggested cleaning method is to (1) place a drop of ethyl alcohol (or dilute tissue culture glassware detergent) on the surface of a slide; (2) clean with tissue; and (3) pass the clean slide through a bunsen flame and cool.

(1) Place a loopful of the culture on a clean slide

(4) Fix by passing the slide rapidly through the bunsen flame three times

(2) Spread in a thin film over the slide

(3) Air dry

Figure 12-1   Preparing a smear for staining.

let is more reactive and usually requires only about 10 seconds. Carbol fuchsin is an even faster dye, usually requiring only 5 seconds. Its reactivity is so great that you may have difficulties from overstaining, especially in preparations that contain large amounts of organic material and debris. Carbol fuchsin is a mixture of the basic dye fuchsin and phenol.

1. Place the slides you fixed in the first part of the procedure on the wire staining rack over the staining tray.
2. Flood each of the fixed smears with approximately five drops of one of the dyes and allow time for it to act: methylene blue 30 seconds, crystal violet 10 seconds, carbol fuchsin 5 seconds.
3. Wash the stained preparations with water from the wash bottle.
4. Dry the slides between blotting paper.
5. Examine the stained preparations under the oil-immersion objective of the microscope and make drawings of a good field. It will be evident if you have made too heavy a smear. A heavy film results in a conglomerate mass of stained material with few if any individual cells visible.

Look closely for significant differences in cell size, shape, and arrangement.

A variety of cells will be found in the saliva stain: several forms of bacteria and fungi, epithelial cells, salivary corpuscles, and leucocytes.

The predominant charge on a bacterial cell (or a protein) is a function of the acidity of its environment. Decreasing the acidity (raising pH) increases the net negative charge on the cell, providing a stronger attraction to basic dyes. The reverse holds true for acidic dyes. Therefore, basic dyes stain poorly at a low pH, and acidic dyes stain poorly at high pH.

Illustrate and label your observations on the report sheet.

QUESTIONS

1. Was the organism isolated from the subsurface of an agar medium similar in cultural and microscopic appearance to either of the isolates?
2. Can you tell whether all of your isolates were pure cultures?

REPORT 12

Direct Staining with Basic Dyes

Name _____

Desk No. _____

Illustrate and label. Drawings of microscopic appearance should illustrate differences in cell size, shape, and arrangement between the isolates.

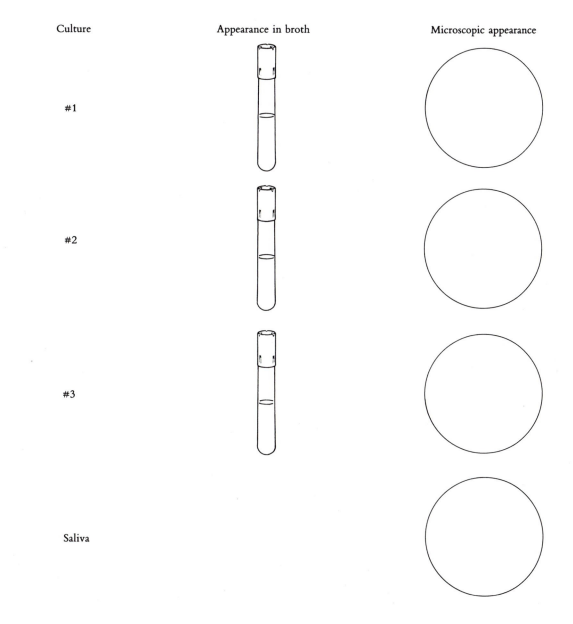

Culture	Appearance in broth	Microscopic appearance
#1		
#2		
#3		
Saliva		

DIFFERENTIAL STAINING TECHNIQUES Simple staining works because bacterial cells differ chemically from their surroundings and thus can be stained to contrast with their environment. Microorganisms also differ chemically and physically from one another, and thus react differently to a staining procedure. This is the basic principle on which differential staining depends. Thus, we can define **differential staining** as a method of distinguishing between types of bacteria.

13. *Gram Stain*

The **gram stain**, the most useful bacteriological staining procedure, is a differential stain. This procedure divides bacteria into two groups: gram-positive and gram-negative. The difference between the two types of cells is thought to be due to variance in their surface layers or walls. In general, gram-positive and gram-negative organisms tend to react differently with many physical and chemical agents.

The gram stain requires four solutions: a basic dye, a mordant, a decolorizing agent, and a counterstain.

Basic dyes have been discussed previously. A **mordant** is a substance that increases the affinity between the cell and the dye, that is, it helps fix the dye on the cell in some way. Examples of mordants are acids, bases, metallic salts, and iodine. Under the action of a mordant, a cell is more intensely stained; it is also much more difficult to wash out the stain after the application of a mordant.

A **decolorizing agent**, as the name implies, is a substance that removes the dye from a stained cell. Some stained cells decolorize more easily than others, and it is this variation in the rate of decolorization that differentiates diverse types of bacteria when the gram stain and other differential stains are used.

The **counterstain** is a basic dye of a different color from the initial one. The purpose of the counterstain is to give the decolorized cells a color that is different from that of the initial stain. Those organisms that are not easily decolorized retain the color of the initial basic stain, and those that are easily decolorized take the color of the counterstain.

Let us now summarize gram staining. The first step consists of staining the cells intensely with a basic dye. This is followed by a treatment of these stained cells with a mordant, for example, iodine. The cells are then treated with a decolorizing agent, such as alcohol. The cells that retain the basic dye after decolorization are called **gram-positive**, and those that are decolorized are **gram-negative** and can be restained with a counterstain. (For illustrations of the steps of this procedure, see Figure 13-1.)

PROCEDURE

Use young cultures (18–24 hours) in which differences in cell-wall structures are retained; old cultures tend to lose gram-positiveness.

1. Prepare smears of *Bacillus cereus, Streptococcus faecalis,* and one mixed smear of *Escherichia coli* and *Micrococcus luteus.* Fix these preparations with heat.
2. Stain with crystal violet for about 30 seconds.
3. Rinse with water.
4. Cover the film with Gram's iodine and allow it to act for about 30 seconds.
5. Rinse with water.

(1) Stain with crystal violet for 30 seconds

(2) Wash off stain with water

(3) Cover with iodine for 30 seconds

(4) Rinse with water

(5) Decolorize with 95% alcohol for 10–20 seconds and rinse with water

(6) Counterstain with safranin for 30 seconds

(7) Rinse with water

(8) Blot off excess water

Figure 13-1 Gram-staining procedure.

Table 13-1   Steps in the Gram Stains

		Results	
Step	*Procedure*	*Gram +*	*Gram −*
Initial stain	Crystal violet for 30 seconds	Stains purple	Stains purple
Mordant	Iodine for 30 seconds	Remains purple	Remains purple
Decolorization	95% ethanol for 10–20 seconds	Remains purple	Becomes colorless
Counterstain	Safranin for 20–30 seconds	Remains purple	Stains pink

6. Decolorize with 95% alcohol. For a thin smear, 10–20 seconds is long enough; after the proper time interval, alcohol drippings from the slide are no longer colored.
7. Rinse with water.
8. Counterstain with safranin for 20–30 seconds.
9. Rinse with water and blot dry.
10. Examine under the oil-immersion objective.
11. Make drawings on the report sheet.

What happens in each step of this procedure? If we were to examine gram-positive and gram-negative cells after each step, we would observe the results outlined in Table 13-1.

The gram stain differentiates bacteria into two groups because of differences in their cell walls. The gram-positive wall in most bacteria is primarily a thick layer of peptidoglycan, and the gram-negative wall is a multilayered structure composed of a thin layer of peptidoglycan surrounded by an outer layer of protein and lipopolysaccharide. The differing permeability of these surface structures to the reagents of the gram stain accounts for the stain's ability to differentiate the two types of organisms. For a more detailed explanation of the gram reaction, consult a microbiology text or a review article on microbial cytology.

Demonstration by the Instructor

Your instructor will demonstrate slides that were prepared with different modifications of the gram procedure. Observe a demonstration of the effects of variations in staining procedure on the gram reaction.

QUESTIONS

1. Can you vary the primary stain and counterstain?
2. What is the effect of replacing iodine with other oxidizing agents?
3. How well do alcohols other than ethyl serve as decolorizing agents?
4. What is the influence of pH on the gram reaction?

Name _____

Desk No. _____

Make drawings of your observations, using appropriately colored pencils to indicate the gram reaction of each culture.

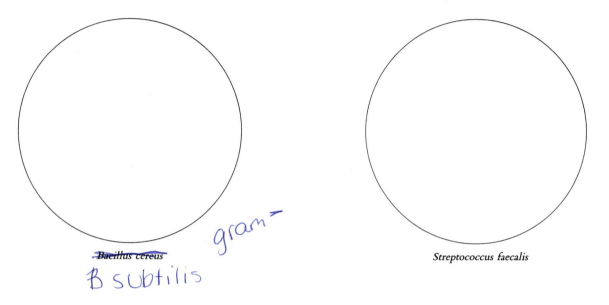

~~Bacillus cereus~~

B subtilis

gram –

Streptococcus faecalis

~~Mixed smear of~~ *Escherichia coli* ~~and Micrococcus luteus~~

gram +

14. *Acid-Fast Staining*

The **acid-fast stain** is a differential stain that measures the resistance of stained cells to decolorization by acids. The property of acid fastness in certain mycobacteria and actinomycetes is correlated with their high lipid content. Staining these bacteria requires heating with dyes of strong affinity for such cells. Once stained they are difficult to decolorize.

This procedure entails staining with hot carbol fuchsin, decolorizing with an acid–alcohol solution, and counterstaining. The acid-fast bacteria are more slowly decolorized by the acid–alcohol solution than other microorganisms and retain the color of the initial stain.

This stain is used primarily in the diagnosis and study of diseases caused by acid-fast species, such as tuberculosis and leprosy. It also serves as a differential stain for a number of harmless acid-fast saprophytes.

PROCEDURE

1. Prepare a smear of combined *Mycobacterium* sp. and *Staphylococcus epidermidis* on a clean slide. Try to break up clumps with your loop to obtain isolated cells. Spread the mixture until the smear is about the size of a quarter.
2. Allow the smear to dry in air. Do *not* shake the slide to hurry up the process of drying. When the slide is completely dry, not sooner, fix it with heat.
3. Bring about ½ inch of water to boil in a cup, hook a wire platform over the edges of the cup, and place the slide on the platform.
4. Stain the smear for 10 minutes with carbol fuchsin by placing a small piece of blotting paper or toweling over the smear and flooding it with the dye. *Caution:* Do not allow the smear to dry during the staining process. Avoid flooding the preparation with too

much dye since it will drop and splash in the steam bath.
5. Wash with water.
6. Decolorize in acid alcohol (95% ethyl alcohol containing 2.5% HNO_3) for 10–30 seconds.
7. Wash with water.
8. Counterstain with methylene blue for 30 seconds (without heating).
9. Wash with water and blot dry.
10. Examine under the oil-immersion objective.
11. Make drawings on the report sheet.

Demonstration by the Instructor

Observe the demonstration of the acid-fast stain in which the primary and counterstains have been varied and in which the heating step has been omitted.

Additional or Optional Preparation

Using the acid-fast staining techniques, stain some tubercular sputum that has been autoclaved. Prepare the smear by crushing and smearing the sputum in a thin film between two slides. After staining, look for the presence of acid-fast organisms, noting how they differ in appearance from other organisms in the sputum. Acid-fast organisms are not decolorized by the acid–alcohol and appear red. Other organisms are stained blue (Color Plate 1).

QUESTIONS

1. What is the effect of varying the primary stain and counterstain? Of omitting heating?
2. Are acid-fast organisms gram-positive or gram-negative?

Acid-Fast Staining

Name _____

Desk No. _____

Make drawings of your observations.

Mixture of *Mycobacterium*
and *Escherichia*

Tubercular sputum

15. *Structural Stains*

Structural stains provide the contrast needed to see various structures of bacterial cells with the light microscope.

ENDOSPORE STAIN

Species of the genera *Bacillus* and *Clostridium* produce a structure called the **endospore**. Unlike the vegetative cell that produces it, the endospore is a resistant body, capable of surviving for long periods in environments made unfavorable by high temperatures or toxic chemicals. The endospore resists staining and, once stained, strongly resists decolorization and counterstaining. The staining method given here uses this characteristic for the microscopic observation of endospores (Figure 15-1).

The *malachite green (Schaeffer and Fulton) stain* employs hot malachite green, an intense stain that is not removed from the endospore by washing. Safranin is the counterstain. Thus the endospore stains green, but the rest of the cell (and cells without endospores) stains red (Color Plate 2).

PROCEDURE

1. Prepare smears of *B. cereus* or *C. sporogenes* cultures, dry the smears in air, and fix them with heat.
2. Place the slides on a staining rack above boiling water.
3. Cover the smears with small pieces of paper towel, keep saturated with malachite green (5% aqueous solution), and continue heating for 5 minutes.
4. Wash gently with water.
5. Counterstain with safranin for 30 seconds.
6. Wash with water and blot dry.
7. Examine under the oil-immersion objective.
8. Make drawings.

The phase microscope (see Exercise 4) is effective in observing the endospore without staining; it appears as a dense white structure in the cell (Figure 15-2). If a phase microscope is available, observe the unstained endospores in cultures of *B. cereus* and *C. sporogenes*.

CAPSULE STAIN

Many bacteria are enveloped in a gummy layer of varying thickness called the capsule, which stains less intensely than the cell. Because the capsule often appears as an unstained area around a stained cell, it can be confused with unstained artifacts such as empty zones that result from shrinkage of cells. The best capsule stains are those that contrast the capsule with the cell and its environment (Figure 15-3).

PROCEDURE

1. Prepare a turbid suspension of *K. pneumoniae* by transferring several loopfuls of growth from the agar slant to a tube containing a small amount of broth.
2. Place a drop (5 or 6 loopfuls) of the Congo red solution on a glass slide.
3. Remove a loopful of the bacterial suspension, mix with the Congo red, and spread gently over the glass slide. Allow to dry at room temperature. **Do not heat-fix!**
4. Flood the smear with Maneval's stain, permitting it to remain in contact for at least 1 minute.
5. Wash with tap water and blot dry.
6. Examine under the oil-immersion objective. The capsules are unstained, the background is blue or black, and the bacterial cells are stained red to reddish brown, depending on the thickness of the preparation.
7. Make drawings on the report sheet.

Demonstration by the Instruction — Flagella Stain

The diameters of the flagella of bacteria are below the limits of the resolving power of the light microscope. Direct observation of these structures

(1) Place piece of paper towel on heat–fixed smear

(4) Rinse with water

(2) Add malachite green

(5) Add safranin

(6) Rinse with water

(3) Remove from heat and discard paper

(7) Blot dry

Figure 15-1 Steps in the endospore stain.

Figure 15-2 Electron micrograph of an endospore of *Bacillus cereus.* (From G. B. Chapman, *J. Bact. 71*:348, 1956.)

Figure 15-3 Capsule stain under the light microscope. (From W. H. Taylor, Jr., and E. Juni, *J. Bact. 81*:698, 1961.)

was, therefore, impossible before the electron microscope came into use. Special staining techniques, however, cover the surface of the flagella with a mordant and increase their apparent diameter, bringing them within the limits of resolution of the light microscope.

Observe the demonstration of various flagellar arrangements under the light microscope (Color Plate 3).

QUESTIONS

1. Can you detect differences in the relative locations, sizes, and shapes of the endospores of *B. cereus* and *C. sporogenes*?

2. How would an endospore-containing cell of *B. cereus* appear after gram staining? (Check your drawings from Exercise 13.)
3. What is the importance of the cell wall to the cell?
4. How do the cell walls of bacteria differ from those of plants?
5. Why doesn't the whole cell stain red with the cell-wall stain?
6. What is the role of the capsule in the economy of the cell?
7. What is the importance of the capsule in pathology? In the food industry?

Structural Stains

Name _____

Desk No. _____

ENDOSPORE STAIN

Make labeled drawings of your observations, using appropriately colored pencils to differentiate vegetative cells and endospores.

Malachite green stain

CAPSULE STAIN

Make labeled drawings of your observations, using colored pencils to differentiate the cells and capsules.

Determining Numbers of Microbes

Practically every phase of microbiology requires methods for measuring microbial numbers. Either cell numbers or cell masses are measured. Methods that measure cell numbers are primarily useful in enumerating unicellular organisms such as bacteria and yeasts; the measurement of cell masses can be employed for all microbial types, including the long filamentous types, such as the molds, that cannot be enumerated by measuring cell numbers.

The most common method of measuring cell numbers is the **plate** or **colony count**, which is based on the theoretical assumption that one bacterial cell or clump of cells gives rise to one colony and, therefore, that the number of colonies that develop on an inoculated agar plate corresponds to the original bacterial count. This method will be discussed in Exercise 17.

In the **dilution count**, a series of tubes of liquid medium are inoculated with the dilutions of a serial dilution of a sample. After incubation, all tubes that contain cells that have grown will be more or less turbid, depending on the number of cells in the dilution, and some tubes of the higher dilutions may be clear, that is, the inocula did not contain any viable bacteria. The population is estimated from the highest dilution that shows growth. In Exercise 57 you will use a modification of the dilution count to determine bacterial numbers in water samples.

Colony counts can be made using the **membrane filter**. By passing a sample through a bacteria-retaining membrane filter, you can trap the cells on the surface. A suitable growth medium is then introduced, the trapped cells grow on the surface of the filter, and the number of colonies indicates the extent of the bacterial population. The use of the membrane filter in the determination of fecal bacteria in water will also be described in Exercise 57.

These culture methods measure the numbers of viable cells that grow in the media. There are several methods that directly measure all cells, living or dead, in a sample. These direct methods are not as precise as culture methods, but they are less time-consuming. One of the most common is the **direct microscopic count**, in which a measured volume of a sample is spread over a measured area of a slide. By counting microbes in a measured volume and multiplying by an appropriate factor, you can calculate the number of organisms in the whole sample. To make the procedure easier, there are special counting chambers (for example, the Petroff–Hauser slide) that have depressions of a known depth and volume, marked off into squared areas. The organisms in an area (for example, 50 or 100 squares) are counted, and the total number of bacteria in the sample can be calculated from the proportion of the total volume known to be contained in the chamber.

In the **Breed method**, a known volume of sample is spread over 1 square centimeter of a slide and is dried, fixed, and stained. The numbers of microorganisms in a large number of microscopic fields are then determined. Since the area of the microscopic field is known, the bacterial count can be calculated.

Cells or particles can also be counted electronically. With the **Coulter counter**, a sample is passed through a small orifice, and from the measured change in electrical resistance the number of particles and their size can be determined. However, since this method measures any particle in a sample, it is applicable only to aqueous suspensions of microorganisms and cannot be used to determine numbers of microorganisms in soil, for example.

Measuring the cell masses in a culture estimates the total cellular protoplasm per milliliter of culture. The more common methods for determining cell masses are light-scattering or turbidimetric methods, chemical estimates of cellular components, dry-weight methods, and cell-volume methods.

Turbidimetric methods, used to measure cell masses, depend on the fact that cells in a liquid medium block or scatter light in proportion to their total mass in the culture. These methods will be discussed and illustrated in Exercise 19.

Chemical estimates of cell masses are made by measuring the amount of some chemical component of the cell, for example, cellular nitrogen, protein, phosphorus, or ATP. Under standardized conditions, the amount of any component present provides means for a reasonable accurate determination of the total biomass in a culture.

The **dry weight** of cells or mycelia from a given volume of growth medium can be determined as a measure of cell mass. For such measurements, the culture is harvested by either centrifugation or filtration, the cells are washed, and their weight is determined after drying. This method is used in Exercise 26.

Cell volume is determined by placing a standard quantity of liquid culture in a calibrated centrifuge tube and measuring the volume of the wet pellet after centrifugation.

16. *Pipetting and Dilution*

Students who take introductory courses in microbiology come to them with various science backgrounds. Some of you are familiar with quantitative pipetting and calculating dilutions, and others are not. The purpose of this experiment is to familiarize you with the materials and techniques of making dilutions of liquids to prepare you, if you are not already prepared, to make the quantitative dilutions that are so important in microbiology.

PIPET FAMILIARIZATION

Many types of pipets are used in microbiology. Some are calibrated to deliver the last drop of fluid ("blow out," Figure 16–1C–G; note double band near neck of pipet) and others are not (Figure 16-1A, B). Your instructor will let you examine the pipets you will use in your course. Make sure you understand how to read the markings on these pipets.

DYE DILUTION

PROCEDURE

Using the materials provided, you will prepare dilutions of dye solution and present these to your instructor for evaluation and discussion (Figure 16-2).

1. With a 10 ml pipet, prepare a dilution blank of 4.5 ml of distilled water in a 16 × 150 mm tube. With the same pipet, prepare a 9 ml dilution blank in another tube. Label the tubes.
2. With the same 10 ml pipet, prepare one dilution blank of 9.9 ml of distilled water in a 16 × 150 mm tube; label the tube.
3. With the same 10 ml pipet, prepare one dilution blank of 4.0 ml and one dilution

Figure 16-1 Types of pipets. *A* and *B*: Measuring pipets. *C* and *D*: Serological pipets. *E, F,* and *G*: Bacteriological pipets. *H*: Volac pipet controller. *I*: Pipet pump, filler/dispenser (rotation of wheel counter-clockwise sucks up fluid; rotation clockwise dispenses fluid).

Figure 16-2   Dilution scheme.

blank of 9.5 ml of distilled water in $16 \times$ 150 mm tubes; label the tubes.

4. With the same 10 ml pipet, prepare one dilution blank of 3.0 ml and one dilution blank of 12.0 ml of distilled water in $16 \times$ 150 mm tubes; label the tubes.

5. With a 1 ml pipet, make two serial 10-fold $(1:10)$ dilutions of the dye solution provided in the 4.5 and 9 ml dilution blanks. Mix the solutions as you deliver by drawing the pipet full to about 1 ml and expelling against the side of the tube just above the meniscus as you withdraw the pipet; repeat, mixing 4 to 7 times. (This mixing procedure is called *serological mixing* and is not used with bacterial cultures to avoid aerosols.)

   Note: If this were a plate counting exercise:
   (a) Pipets and tubes would have to be handled aseptically.
   (b) Fresh pipets would have to be used for each delivery because the bacterial cells are particulate, and error would result from picking up cells from the inside of the pipet as the pipet was filled for the next transfer.

6. Using serological mixing and a 0.1 ml pipet, make a 100-fold $(1:100)$ dilution of the original dye solution into three 9.9 ml dilution blanks.

7. Using serological mixing and a fresh 1 ml pipet, make a 5-fold $(1:5)$ and then a 20-fold $(1:20)$ dilution of the original dye solution into the 4.0 ml dilution blank and the 9.5 ml dilution blank, respectively.

8. Using serological mixing and a fresh 1 ml pipet, make a 4-fold $(1:4)$ and then a 25-fold $(1:25)$ dilution of the original dye solution into the 3.0 ml dilution blank and then the 12.0 ml dilution blank.

9. The last tube of each dilution series should appear the same. Be sure you understand why.

A GUIDE TO SOLVING DILUTION PROBLEMS

By definition, a dilution factor is the ratio of the final concentration to the initial concentration of a diluted sample. Avoiding theory and resorting to "plug in and grind out" equations, we find three formulas to be of great help.

Equation 1:

$$\text{dilution factor} = \frac{\text{cells/ml AFTER dilution}}{\text{cells/ml BEFORE dilution}}$$

If the number of cells is not known but only the volumes, Equation 2 should be used.

Equation 2:

$$\text{dilution factor} = \frac{\text{ml of sample}}{\text{ml of sample added} + \text{ml to which it was added}}$$

Equation 3:

$$\text{total dilution factor} = \frac{\text{product of all intermediate}}{\text{dilution factors}}$$

Any dilution problem can be solved by applying these three equations. Many methods work, but one generalized form is outlined here.

Step 1. Find the final concentration (cells/ml). This is often given in the problem.
Step 2. Calculate the final dilution factor for each step (Equation 1 or 2).
Step 3. Calculate the total dilution factor (Equation 3).
Step 4. Calculate the initial concentration.

Sample Problem. A pond is to be sampled for water purity. A 10 ml sample of pond water is analyzed; 2.5 ml of the sample is added to 97.5 ml of pure water; 1.6 ml of this is removed and added to 14.4 ml of pure water; 0.1 ml of this is added to 99.9 ml; 0.1 ml of this is then plated. After 24 hours, 200 colonies were counted. What is the concentration of bacteria in the pond?

Step 1. Final concentration of cells/ml:

$$\frac{200 \text{ cells}}{0.1 \text{ ml}} = 2000 \text{ cells/ml}$$

Step 2. Dilution factor for each step:

(a) $2.5/(2.5 + 97.5) = 2.5 \times 10^{-2}$
(b) $1.6/(1.6 + 14.4) = 10^{-1}$
(c) $0.1/(0.1 + 99.9) = 10^{-3}$

Step 3. Total dilution factor:

$$(2.5 \times 10^{-2})(10^{-1})(10^{-3}) = 2.5 \times 10^{-6}$$

Step 4. Initial concentration:

$$2.5 \times 10^{-6} = 2000/??? $$
$$= 8 \times 10^{8} \text{ cell/ml}$$

A typical protocol for use with spread plates is shown in Figure 16-3.

Figure 16-3 Sample dilution protocol for use with spread plates.

17. *Quantitative Plating Methods*

Because one viable bacterial cell or group of cells (colony-forming unit) gives rise to one colony, you can also use plating to determine the number of bacteria in a material. To secure a bacterial count, the sample is diluted and plated, and after incubation the colonies that develop are counted. The bacterial count of the original sample is then determined by multiplying the number of colonies that develop by the degree of dilution (dilution factor) of the sample in the plate. Dilutions are usually expressed as negative exponents, for example, 10^{-5} is used instead of 1/100,000.

Two main quantitative techniques are in common use. The **spread plate** method distributes the sample over the surface of nutrient agar plate, and colonies are counted after incubation. The **pour plate** method ("plating out" in scientific jargon) introduces the sample into a tube of melted, warm (45°C) agar culture medium which is mixed thoroughly and is then poured into a petri plate. Colonies are counted after incubation. In both techniques, dilutions of the sample are prepared and plated. In this exercise you will have the opportunity to use one of these methods.

To dilute the sample quantitatively, a 1 g or a 1 ml portion is usually diluted stepwise through a series of tubes or bottles containing a known amount of sterile buffer solution.*

Any plating procedure is simplified by arranging your petri plates and dilution tubes at the beginning. Mark the plates and dilution tubes according to the dilutions they will hold. Label the plates with such other marks of identification as may be needed.

SPREAD PLATE

In this exercise you will add a 1 ml portion of a water sample to a dilution blank containing 9 ml of sterile diluent, resulting in a 10-fold dilution of

your sample. By continuing this dilution stepwise through additional dilution tubes, you can dilute the sample 10^{-2}, 10^{-3}, and more, in the manner outlined in Figure 16-3. Be sure to change pipets with each dilution tube. In the diagram, the transfers that can be made with the same pipet are marked with the same letter. Amounts of 1/10 ml from each dilution tube are spread on the surfaces of agar plates (Figure 17-1), and after incubation the colonies that arise are counted to determine the number of cells at that dilution. You can then use the dilution factor to determine the number of organisms in the original material.

PROCEDURE

1. Obtain three dilution tubes and four nutrient agar plates. Label the tubes 10^{-1}, 10^{-2}, and 10^{-3} and the plates 10^{-1} to 10^{-4}.

2. With a sterile pipet remove 1.1 ml of the water sample. Pipet 0.1 ml on the agar surface of plate 10^{-1} and the remaining 1.0 ml to the 10^{-1} dilution tube. This tube now contains 1 ml of the original sample diluted 10 times; thus, 1.0 ml of this dilution is equivalent to 0.1 ml of the original sample. Mix thoroughly with a vortex mixer (Figure 9-6) or by vigorously shaking (Figure 17-5).

3. With another sterile pipet, remove 1.1 ml from this dilution tube (the 10^{-1} dilution) and place 0.1 ml on the agar surface of plate 10^{-2}. Pipet the remaining 1.0 ml into another sterile dilution tube (the 10^{-2} dilution). Vigorously mix this dilution. Using another sterile pipet, remove 1.1 ml from the 10^{-2} dilution tube, place 0.1 ml on the agar surface of the plate 10^{-3}, and pipet the remaining 1.0 ml into a third sterile dilution tube (the 10^{-3} dilution). After vigorously mixing the final dilution, transfer 0.1 ml to the agar surface of plate 10^{-4} with a sterile pipet.

4. Spread the sample on each plate using a sterile glass spreader as illustrated in Figure 17-2. In this technique the glass spreader is

---

*A recommended diluent is phosphate buffer solution (34 g KH_2PO_4 in 500 ml H_2O; adjust pH to 7.2 with 1 N NaOH; bring total volume to 1,000 ml). Dilute 1 ml to 800 ml just before use.

Figure 17-1 Pipetting into a petri plate.

Dip spreader
in alcohol

ETHANOL

Petri dish
to smother flame
in case of accident

Light
the spreader

Let spreader
burn & then cool

Spreading
the plate

Turntable

Figure 17-2 Steps in spreading a plate.

sterilized by dipping it in alcohol, and then the alcohol is burned off. Allow the spreader to cool. *Caution:* It must be thoroughly cooled. Spread the sample over the surface of the agar plate by revolving the plate on your desk or on a turntable.

5. Incubate the petri plates at 30°C in an inverted position until the next laboratory.

6. Count the colonies on each plate that contains between 20 and 200 colonies.

Counting Plates

Colonies that develop in agar plates can be effectively counted in a device such as the Quebec colony counter, which provides an indirect transmitted light source and a magnifying glass (Figure 17-3). Invert the plate on the counter. Mark each colony on the glass with a pen or marking pencil to avoid recounting. If many plates are to be counted, a tally is helpful.*

Colonies known as "spreaders" often develop in moisture films on the surface of the agar or between the agar and the bottom of the petri plate. If spreaders are large and cover other colonies, discard the plates. Many spreaders are small, grayish, and filmy. If they are small and discrete, they can each be counted as a single colony.

_____

*Electronic counters are used when large numbers of plates are routinely counted.

Figure 17-3   Quebec colony counter.

Calculation of Count

The number of bacteria or colony-forming units per milliliter of the original sample is obtained by multiplying the number of colonies on the plate by the dilution factor. For example, if you counted 150 colonies on the 10^{-3} dilution plate, you can calculate $150 \times 1000 = 150,000$ bacteria per ml.

It is advisable to prepare duplicate plates of each dilution. The counts of duplicate plates should be averaged to the nearest second-place, significant number. Counts of 148 and 154 on the 10^{-3} plate should be averaged to 151, and the count reported as 1.5×10^5 bacteria per ml.

POUR PLATE

In this exercise you are to determine the microbe count of two water samples A and Z. A is known to be of good quality, containing a low bacterial count. Z is an impure sample that contains large numbers of bacteria, and it must be diluted to obtain countable plates. For this dilution you will add a 1.0 ml portion of water sample to a blank containing 99 ml of sterile diluent, resulting in a 100-fold dilution of your sample. By continuing this dilution stepwise through additional dilution bottles, you can dilute the sample 10^{-4}, 10^{-6}, and more. Then, by plating out 1 ml and 0.1 ml portions from each dilution blank (Figure 17-4), you can plate the sample at 10^{-1}, 10^{-2}, 10^{-3}, 10^{-4}, and higher dilutions.

PROCEDURE

1. Melt five tubes of nutrient agar (15 ml each). Cool to 45°C and have these tubes ready to use when the dilutions are made. No more than a few minutes should pass between placing the sample in the petri plates and mixing the sample with agar.

2. With a sterile pipette remove 1.1 ml of sample A, and place 0.1 ml into a sterile petri plate (Figure 17-1). Drain the remaining 1.0 ml of the same sample into another sterile petri plate. Mark the first plate A-10^{-1} and the second A-1.

3. Sample Z, containing larger numbers of bacteria, cannot be plated out directly; it must be diluted. With a sterile pipet, remove 1 ml of sample Z and put it into a 99-ml dilution

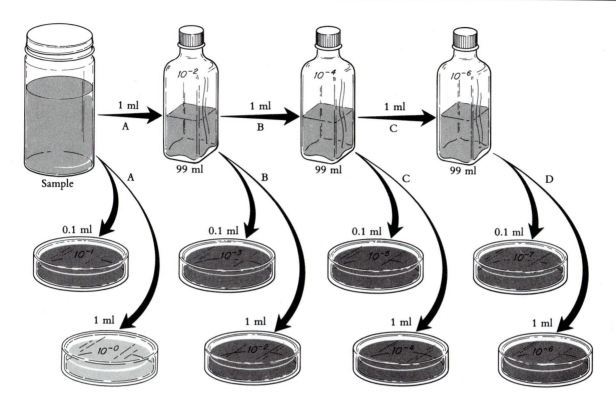

Figure 17-4 Diluting the sample for the pour plate quantitative plating method.

blank. This bottle now contains 1 ml of the original sample diluted 100 times; thus, 1.0 ml of this dilution is equivalent to 0.01 ml of the original sample.

4. Shake the bottle vigorously, 25 times, moving your forearm over an arc of 15 inches or so to insure a uniform mixing of the sample and to break apart bacteria that may be clumping (Figure 17-5).

5. With another sterile pipet, remove 1.1 ml from this bottle (the 10^{-2} dilution) and place 0.1 ml in a sterile petri plate (a 10^{-3} dilution of the original); mark it Z-10^{-3}. Place the remaining 1.0 ml in another sterile petri plate (a 10^{-2} dilution of the original); mark it Z-10^{-2}.

6. With the same pipet take another 1.0-ml aliquot from the 1/100 dilution bottle and transfer it to another 99-ml sterile dilution blank. Shake this bottle 25 times as before, and transfer 1.0 ml of the mixed sample to a petri plate. Label this plate Z-10^{-4}.

7. Pour the melted, cooled agar into the petri plates and mix the sample portions thoroughly with the agar. One of the best ways

Figure 17-5 Mixing a sample in a water blank.

of mixing is to tilt a plate slightly so that a wave of agar and sample travels around the plate a dozen or so times. Take care not to slop the agar over the edge of the plate.

8. Set the petri plates on a level surface. When cool, incubate at 30°C in an inverted position.

9. Count the colonies on each plate that contains between 30 and 300 colonies.

10. Complete the charts on the report sheet.

Refer to the previous section on spread plates for information on colony counting and calculation of the plate count.

QUESTIONS

1. Why do you count plates with 20–200 colonies in the spread-plate technique and plates with 30–300 colonies in the pour-plate technique?

2. Are colonies likely to arise from clumps of bacteria rather than from single cells? What is meant by the term *colony-forming unit* (CFU)?

3. Why must pipets be changed between dilutions?

Quantitative Plating Methods

Complete these charts.

Spread plate

	Dilution counted	Number of colonies	Plate count per ml of original sample
Sample A		1	

Pour plate

	Dilution counted	Number of colonies	Plate count per ml of original sample
Sample A		3	
Sample Z		1	

2

18. *Most Probable Number*

One fast quantitative method for routinely examining water for fecal organisms is the most probable number (MPN) technique. Adapted as a standard method for the examination of drinking water and waste water by the American Public Health Association, it is basically a statistical approach to the quantitative estimation of bacterial numbers. Samples are serially diluted to the point at which there are few or no viable microorganisms. The detection of this end point is based on multiple serial dilutions that are inoculated into a suitable growth medium. Statistical tables are available to estimate the size of bacterial populations based on the number of replicate tubes (3, 5, or 10) of each dilution. The MPN method has the advantage of being a bit

faster for the microbiologist to perform and evaluate. This exercise can be done in concert with Exercises 17, 19, 57, and 58.

PROCEDURE

1. Add 1 ml of the water sample suspected of containing *E. coli* to a 99 ml water dilution bottle (Figure 18-1). This is your dilution 1 bottle (1×10^{-2}).
2. Shake the bottle vigorously, and aseptically transfer 1 ml of dilution to a second 99 ml dilution bottle (1×10^{-4}). Shake the bottle vigorously.

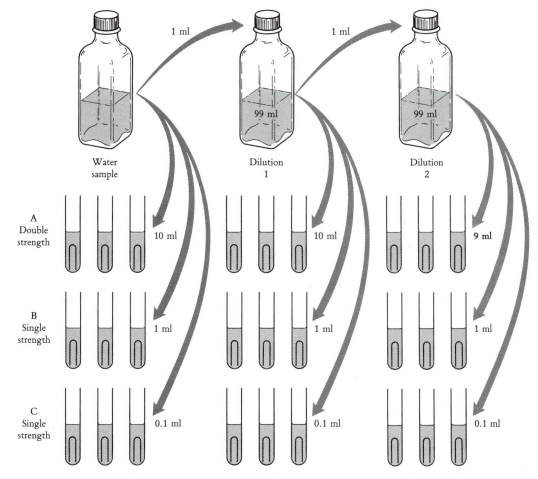

Figure 18-1 Flow chart for the inoculation of media for the MPN technique.

3. For each of the three bottles (original sample, dilution 1, and dilution 2), inoculate tubes as follows:

   (a) Add 10 ml from each bottle to each of three tubes of double-strength phenol red lactose broth with Durham tubes.

   (b) Add 1 ml from each bottle to each of three tubes of single-strength phenol red lactose broth with Durham tubes.

   (c) Add 0.1 ml from each bottle to each of three tubes of single-strength phenol red lactose broth with Durham tubes.

4. Incubate the tubes at 37°C for 24 hours.

5. After incubation, examine the tubes for acid and gas production. Record your results and then select the dilution in which the tubes are neither all positive nor all negative. Using the series with mixed results, compare with the MPN table (Table 18-1) and record the most probable number of coliforms in your sample. Multiply by the dilution factor if necessary.

QUESTIONS

1. What are the advantages and disadvantages of the MPN method compared with the plate-count method?

2. What are some possible mistakes you could make in the MPN method that might affect its accuracy?

3. How would you change the method if you wanted to detect the total numbers of viable bacteria in a sample instead of just *E. coli* and other lactose-fermenting bacteria?

Table 18-1   MPN Determination from Multiple Tube Test

Number of Tubes Giving Positive Reaction			MPN Index per 100 ml	95 Percent Confidence Limits	
3 of 10 ml each	3 of 1 ml each	3 of 0.1 ml each		Lower	Upper
0	0	1	3	<0.5	9
0	1	0	3	<0.5	13
1	0	0	4	<0.5	20
1	0	1	7	1	21
1	1	0	7	1	23
1	1	1	11	3	36
1	2	0	11	3	36
2	0	0	9	1	36
2	0	1	14	3	37
2	1	0	15	3	44
2	1	1	20	7	89
2	2	0	21	4	47
2	2	1	28	10	150
3	0	0	23	4	120
3	0	1	39	7	130
3	0	2	64	15	380
3	1	0	43	7	210
3	1	1	75	14	230
3	1	2	120	30	380
3	2	0	93	15	380
3	2	1	150	30	440
3	2	2	210	35	470
3	3	0	240	36	1300
3	3	1	460	71	2400
3	3	2	1100	150	4800

From: Standard Methods for the Examination of Water and Wastewater, Twelfth edition. (New York: The American Public Health Association, Inc., p. 608.)

Most Probable Number

Name _____

Desk No. _____

	Water Sample			Dilution 1 1:100			Dilution 1:10,000		
	Tube 1	Tube 2	Tube 3	Tube 1	Tube 2	Tube 3	Tube 1	Tube 2	Tube 3
A Double Strength									
B Single Strength 1 ml									
C Single Strength 0.1 ml									

Most probable number _____

19. *Turbidimetric Estimation of Bacterial Growth*

A bacterial culture acts like a colloidal suspension, blocking and reflecting light. Within limits the light absorbed or reflected by a bacterial suspension is directly proportional to the concentration of cells in the culture. Thus, by applying **nephelometry**, measurement of the reflection of light rays, or **turbidimetry**, measurement of the percentage of light absorption, to a bacterial suspension, you can estimate the number of cells in it (Figure 19-1).

For such determinations you will use a **photocolorimeter**. This instrument is a source of monochromatic light (that is, light of a single wavelength, usually supplied through a filter that allows only the desired wavelength of light to be transmitted). The light is passed through a bacterial culture, and the amount of light reflected or transmitted is measured by a photoelectric cell wired to a galvanometer (Figure 19-2). Photocolorimeters are usually used as turbidimeters; only rarely are they used as nephelometers.

Ordinarily, it is more useful to express the turbidity as **absorbance**, which is directly proportional to the cell concentration.

Absorbance (A) is a function of the negative log of the percent transmission [−log of galvanometer reading (G)] and is expressed as $2 - \log G$; that is, $A = \log 100 - \log$ of galvanometer reading.

In measuring cell growth by turbidimetry, the turbidity of the bacterial culture is correlated with some other known measure of cell growth such as the total number of bacteria as determined by the quantitative plate count. After a standard curve has been determined, which shows the relationship between absorbance and the number of bacterial cells, it can be used to estimate the number of bacteria in a suspension. This procedure has many applications, including its use in the clinical laboratory to standardize the inoculum for determining antibiotic susceptibility by the Kirby–Bauer method (described in Exercise 71).

Growth of a microbe is an orderly increase in size and number of its components that is usually followed by cell division. Under optimal conditions growth of a culture of microbes is an orderly increasing or doubling of cells and is termed **balanced growth**.

Figure 19-1   Reading the absorbance of a culture in a turbidimeter.

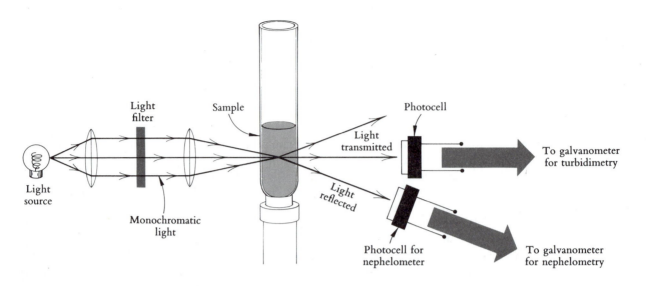

Figure 19-2 Schematic diagram of photocolorimeter that might function as a nephelometer or a turbidimeter.

Microbial growth in laboratory conditions follows a predictable course. After inoculation into a fresh medium, the culture lags in the increase in cell numbers. During this period cells are increasing in size while adapting their synthetic capacity (that is, DNA, RNA, and protein synthesis) for optimal growth. This phase is followed by a period of exponential growth during which the culture is in balanced growth, and the growth rate is constant. In time, the culture exhausts essential nutrients in the environment, and toxic wastes build up, eventually slowing the growth rate. As environmental conditions worsen, the increase in cell numbers stops, and after a time, the cell population begins to decline.

One can follow the growth of a microbial culture by measuring changes in cell mass, viable counts, or any chemical constituent of the cell (for example, RNA or protein). In this experiment you will follow the growth of a microbial culture by measuring two parameters: cell mass and viable count. This gives you the opportunity to compare the two methods.

Figure 19-3 Dilution procedure for pour plating.

Figure 19-4 A nephelometer flask and a Klett–Summerson colorimeter.

PROCEDURE

You will work in pairs to perform this experiment. You will be given a flask of a 12–14 hour culture of *E. coli* growing in nutrient broth. The flask will be in a shaker water bath at 37°C.

1. Immediately after receiving the flask, read and record the absorbance of the culture tube at 650 nanometers (nm) using the photometer or spectrophotometer provided. After taking these absorbance measurements, immediately incubate the culture at 37°C on the shaker provided.

2. Continue taking absorbance readings at 15 minute intervals until the absorbance no longer increases. After each reading, reincubate the culture. Because fluctuations in temperature can slow the growth of this bacterium, incubation of the culture should be interrupted as briefly as possible.

3. To determine viable counts, start plating the culture when the absorbance of the culture exceeds 0.1. The procedure for the plate count is as follows:
   (a) Immediately after reading the culture's absorbance, transfer 1 ml of the culture to a 99 ml dilution blank. Mix well.
   (b) Make two more serial (1 : 100) dilutions. Plate 10^{-5}, 10^{-6}, and 10^{-7} dilutions using nutrient agar (Figure 19-3).

4. Repeat this plate count procedure at 30 minute intervals employing the dilution procedure outlined in Step 3.

5. Incubate all plates at 37°C, and after 24 hours make counts of all samples.

6. Plot the absorbance readings per milliliter versus time on semilog paper provided at the end of this exercise. After your plates have grown, plot the viable counts/ml versus time on the same graph.

7. From your plot of the viable cell count/ml versus time, calculate the mean generation time of *E. coli* during the logarithmic phase of growth.

8. Fill in the chart on the report sheet.

Some instructors may want to modify this experiment. For example, they may want you to measure optical density by working with nephelometer flasks, which are quickly and conveniently inserted into a Klett–Summerson colorimeter (Figure 19-4).

QUESTIONS

1. Why is the galvanometer set at 100% transmission for the uninoculated tube of broth?
2. You will be able to use turbidity as a means of determining cell growth in future experiments. Using the plot of absorbance versus cell count derived in this exercise, you will be able to approximate actual cell counts in future exercises with *Escherichia coli*. Why can't you use this same plot (or standard curve) for other bacteria?
3. What are the advantages of estimating microbial numbers or biomass by this method?
4. What might be some disadvantages?
5. Can you measure all kinds of microbes this way? Why not?

Turbidimetric Estimation
of Bacterial Growth

Name _____

Desk No. _____

Record the following data:

	Time interval (minutes)	A	Plate count	Log of CFU/ml
	0	_____	_____	_____
	15	_____	_____	_____
(Continue readings until A exceeds 0.1)	30	_____	_____	_____
	45	_____	_____	_____
	60	_____	_____	_____
	75	_____	_____	_____
	90	_____	_____	_____
	150	_____	_____	_____

Use the semilog graph paper on the reverse side to plot A and the log of the CFU/ml as functions of time.

Calculation of mean generation time of *E. coli* during log phase of growth:

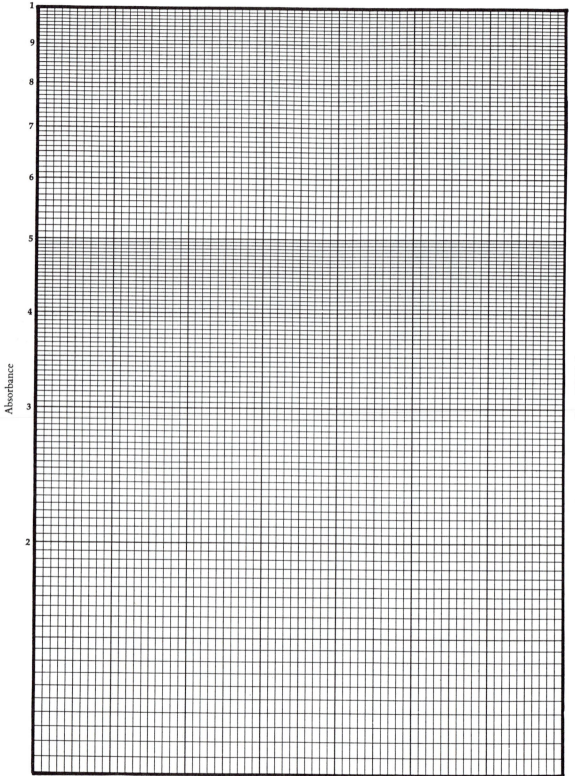

Absorbance

Log CFU/ml

Time (minutes)

Environmental Influences

Environmental effects on microbes fall into three classes: *physical*, the effects of temperatures and pressures; *chemical*, the need for foods and response to poisons; and *biological* (essentially chemical), the influences of coexisting species.

There is an optimum set of environmental conditions for each microbial species, as there is for each plant or animal species, but there is a wider range of tolerance in the microbial kingdom as a whole than in the higher forms of life. The human body, for example, has a rigidly regulated normal temperature of 37°C (98.6°F) that must remain constant in spite of exposure to heat or cold. The bacterial cell has no regulated body temperature but assumes the temperature of its environment. Its response to prolonged exposure to cold or extreme drying is merely to stop enzymatic activities; it does not necessarily die.

Although the vast majority of microbial species require the same conditions that are beneficial to most plant and animal cells, there are "extremists," microorganisms that survive (or even demand) conditions that are immoderate by ordinary standards. From among the variant cells that exist in all large populations, the environment has selected those that are capable of thriving under whatever conditions it has imposed. However, not even the microbes can cope with the two mortal enemies of all life forms: extreme heat and certain chemical poisons, both of which target enzymes and other critical cell constituents, without which life vanishes.

The exercises in this section illustrate the response of various microbial types to environmental conditions. In the fight to control destructive and useful microorganisms, our most powerful tool is the manipulation of the microbe's environment. In our effort to preserve various materials, we have used temperature, osmotic pressure, and oxygen deprivation to check microbial growth. And, as you might expect, the great adaptivity and versatility of microorganisms make their control difficult.

20. *Effects of Temperature on Growth*

Different types of bacteria have distinct requirements for the temperature at which they grow. Between a **maximum temperature**, above which a culture will not develop, and a **minimum temperature**, below which a culture will not develop, is a range in which growth occurs. The most

growth takes place within a limited range called the **optimum temperature**. The optimum temperature for the growth of a particular microbial species is correlated with the temperature of the normal habitat of the organism. For example, the optimum temperature for the growth of organisms that are pathogenic for warm-blooded animals is near that of the blood temperature of the animals, 37°C.

The influence of temperature on microbial growth is actually a reflection of the effect of temperature on the enzymatic reactions in the cells (Figure 20-1). As the temperature is lowered, the enzyme activity, and thus the growth of the cell, is slowed. At the freezing point metabolic activity ceases, not only because of the direct retardation of enzyme activity but also because the cell is deprived of water, which is essential for the uptake of nutrients and the removal of waste products.

As the temperature is raised above the optimum for growth, metabolic activity increases, but at the same time the rate of enzyme and protein breakdown (owing to protein denaturation) also increases, resulting eventually in damage to and then death of the cell.

PROCEDURE

You are provided with 20 tubes of glucose broth.

1. Inoculate one set of five tubes with each of the following cultures: *Escherichia coli, Bacillus stearothermophilus, Pseudomonas fluorescens, Micrococcus luteus.*
2. Incubate one of the five tubes of each culture at each of the following temperatures: 5°, 20°, 35°, 45°, and 55°C. All tubes should be incubated within 5 minutes of inoculation. Those cultures at 40°C or higher should be incubated in well-designed incubators or waterbaths to prevent temperature fluctuation.

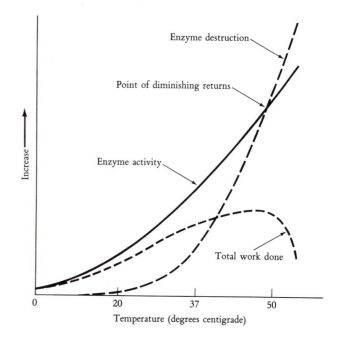

Figure 20-1 Relation of temperature and destruction of enzymes in a mesophilic cell.

3. Observe growth at 2- and 7-day intervals.
4. Record your results on the report sheet.

QUESTIONS

1. What is a thermoduric organism? A psychrotroph? (See Figure 20-2.)
2. From what natural source would you isolate a thermophile? A psychrophile?

Figure 20-2 Temperature growth range of certain groups of bacteria.

REPORT 20

Effects of Temperature on Growth

Name _____

Desk No. _____

Record the results, using a scale from 0 (no growth) to ++++ (very heavy growth).

Organism	Temperature				
	5°C	20°C	35°C	45°C	55°C
Escherichia coli	0	+ +	+ + + +	0	0
Bacillus stearothermophilus	0	+ +	+ + +	0	+
Pseudomonas fluorescens	0	+ + ⊕⊕	⊕⊕ +	⊕ +	⊕ 0
Micrococcus luteus	0	+ + + +	⊕ + +	0	0

21. *Effects of Heat on Vegetative Cells and Spores of Bacteria and on Spores of Yeasts and Molds*

The spores of yeasts and molds differ from bacterial spores in that they are primarily reproductive spores; several to many are produced per yeast cell or mold thallus and they are not heat resistant. In contrast, only one bacterial endospore is produced per cell, and it is notoriously heat-resistant. In this experiment you will compare the heat resistance of various types of microbial spores and vegetative cells.

PROCEDURE

1. You are provided with duplicate tubes of broth cultures of *Bacillus cereus, Escherichia coli, Aspergillus niger,* and *Saccharomyces cerevisiae.* Two tubes of sterile broth are also furnished; inoculate these with a small amount of soil.

2. Immerse one of the duplicate tubes of each set in hot water, carefully regulate at 80°C, for 10 minutes. Be sure that the level of the hot water is well above the level of the tube contents. Remove the tubes from the hot water and cool them immediately in cold water. Incubate both heated and unheated tubes at room temperature until the next laboratory.

3. After incubation, inoculate glucose agar slopes from all tubes. Incubate the slopes at room temperature for 4 days and observe for growth.

4. Make gram stains and spore stains from those slopes that were inoculated from the heated tubes and that show growth.

5. Record your results on the report sheet.

QUESTIONS

1. What conclusions can you draw about the heat resistance of vegetative cells and various types of spores?

2. What characteristics of the bacterial endospore could account for its heat resistance?

3. Is the failure of the heated cultures of yeast and mold to grow on glucose slopes enough evidence to prove that their spores did not survive the treatment?

Effects of Heat on Vegetative Cells and Spores of Bacteria and on Spores of Yeasts and Molds

Name _____

Desk No. _____

Record results, using + to indicate growth and − to indicate the absence of growth.

Slope	Bacillus cereus	Escherichia coli	Aspergillus niger	Saccharomyces cerevisiae	Soil
Heated					
Unheated					

Sketch spore stains from cultures surviving the heat treatment.

22. *Effects of Osmotic Pressure on Growth*

Bacterial cells and the media in which they are suspended have independent osmotic concentrations (a function of substances in solution). When a bacterial cell is placed in a medium, an **osmotic pressure** is exerted across the semipermeable membrane that surrounds the cell.

A microorganism usually grows best in a medium that has a slightly lower osmotic concentration than its own. This causes water to flow into the cell, a condition essential for the diffusion of nutrient substances and for the maintenance of outward pressure (**turgor**). When the osmotic concentration of the medium is considerably lower than that of the cell (a **hypotonic** medium), diffusion of water into the cell is excessive, causing increased turgor. Cells, such as red blood cells, that are not bounded by rigid walls may burst in such a medium, a phenomenon termed **plasmoptysis**. If the osmotic concentration of the medium exceeds that of the cell, the medium is said to be **hypertonic** with respect to the cell. In a hypertonic medium, water leaves the cell, and the cytoplasmic membrane shrinks away from the rigid cell wall, a condition termed **plasmolysis** (Figure 22-1).

This exercise illustrates that microbial species differ markedly in their response to the extremes of osmotic pressure imposed by various environments and that the inhibitory effects of high osmotic concentrations can be used in the preservation of food.

PROCEDURE

1. You are provided with four petri plates of glucose agar containing varying concentrations of NaCl (0.1 *M*, 1 *M*, 3 *M*, 5 *M*) and four petri plates of glucose agar containing varying concentrations of sucrose (0.015 *M*, 0.044 *M*, 0.088 *M*, and 1.75*M*). Label each plate according to its concentration of NaCl or sucrose, and with a wax pencil, divide the bottoms of the plates into six sectors (Figure 22-2). With a loop inoculate each of four sectors of each plate with a different one of the four cultures provided (*Escherichia coli, Saccharomyces cerevisiae, Staphylococcus aureus,* and *Penicillium chrysogenum*). Add some raw hamburger and dried fruit to separate tubes of sterile tap water. Shake vigorously and, with a loop, inoculate the two remaining sectors of each plate with the meat and fruit mixtures. Inoculate only small areas to prevent overgrowth and crowding of the cultures. Label each sector with the name of its culture. Incubate at 30°C and observe for growth at 2- and 7-day intervals.

2. You are also provided with four tubes of glucose broth containing varying concentrations of NaCl (0.1 *M*, 1 *M*, 3 *M*, 5 *M*) and four tubes of glucose broth containing vary-

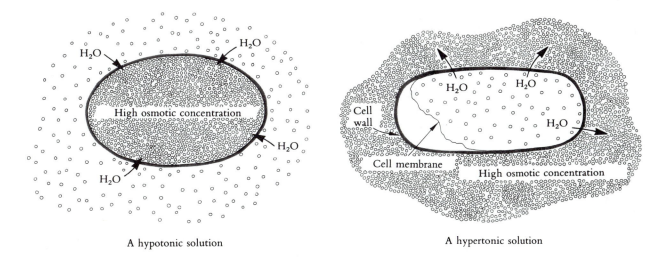

A hypotonic solution   A hypertonic solution

Figure 22-1 Relation of osmotic concentration of the medium to the flow of water into or out of a cell.

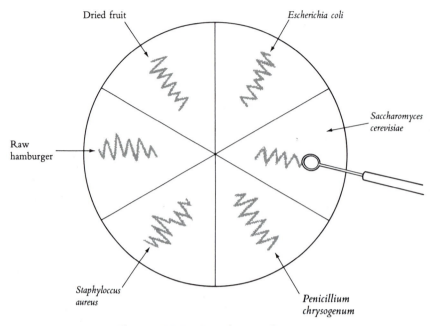

Dried fruit

Escherichia coli

Saccharomyces cerevisiae

Raw hamburger

Staphyloccus aureus

Penicillium chrysogenum

Figure 22-2   Inoculation scheme.

ing concentrations of sucrose (0.015 *M*, 0.044 *M*, 0.088 *M*, and 1.75 *M*). Inoculate each of the NaCl broth tubes with a piece of raw hamburger and each of the sucrose broth tubes with a piece of dried fruit. Incubate at 30°C and observe for growth at 2- and 7-day intervals.

3. Record your results on the report sheet.

Most bacteria are sensitive to high osmotic pressure, in contrast to yeasts and molds which are more tolerant of hypertonic environments. However, some microbial forms, including bacteria, thrive in natural habitats that have high osmotic pressures. If you were to isolate bacteria from the natural salt lakes, you would encounter some species that could tolerate high salt concentrations, and some that would *require* high salt concentrations for optimum growth. The organisms that thrive at high osmotic pressures or in high salt concentrations are referred to as **osmophilic** and **halophilic**, respectively. Organisms that survive but do not thrive under such conditions are referred to as tolerant, **salt-tolerant**, for example.

The effect of osmotic pressure, which is chiefly a drying effect, is of great practical importance in the bacteriology of foods. The preservation of some foods (jam, sauerkraut, condensed milk, salted meats, and fish) is partially or wholly due to their high osmotic concentrations. *Preserved* does not mean that a food is sterile; preserved food may contain many viable forms that will grow readily if osmotic pressure is decreased.

QUESTIONS

1. Are all osmophilic organisms halophilic? Is the reverse true?
2. How is it possible for an organism to grow at high osmotic pressure?

Effects of Osmotic Pressure on Growth

Name _____

Desk No. _____

Record results, using a scale from 0 (no growth) to $++++$ (very heavy growth). Remember that the growth of the bacteria and the yeast will appear quite different from that of the mold.

Additive to medium		Escherichia coli	Saccharomyces cerevisiae	Penicillium chrysogenum	Staphylococcus aureus	Raw hamburger		Dried fruit	
						Plate	Liquid	Plate	Liquid
NaCl	0.1 M								
	1.0 M								
	3.0 M								
	5.0 M								
Sucrose	0.015 M								
	0.044 M								
	0.088 M								
	1.75 M								

23. *Effects of pH of Medium on Growth*

There are optimal hydrogen ion concentrations for most organisms, although they grow over a fairly wide range of pH. Generally, microorganisms grow best at the pH of their natural habitats. For example, many parasites of the human body, such as the pneumococci, grow best at pH values near that of the human blood stream, pH 7.4.

In many cases the microorganism itself, as a result of its metabolism, plays a major role in setting the pH of its environment. Acid-producing bacteria, molds, and yeasts increase the hydrogen ion concentration of their environments and tend to grow best at moderately low pH values. Other bacteria, especially putrefactive forms that decompose proteins into basic amines and ammonia, raise the pH of their environments and thrive in alkaline conditions.

Since pH limits the kinds and types of microorganisms that grow in any given condition, pH control has practical importance. For example, sulfur is sometimes added to soil as a means of inhibiting the growth of the microbes that cause potato scab.

Sulfur is converted to acid by other soil bacteria, and the resulting drop in pH inhibits the potato pathogen.

PROCEDURE

1. You are provided with four glucose-agar plates of varying pH (pH 3, 5, 7, and 9). Designate the pH of each. Mark quadrants with a wax pencil on the bottom of each plate (Figure 23-1). With a loop, inoculate each quadrant with a different one of the cultures provided (*Staphylococcus aureus, Escherichia coli, Saccharomyces cerevisiae,* and *Penicillium chrysogenum*). Do this for all plates. Inoculate only a small area to prevent overgrowth and crowding.
2. Incubate at 30°C.
3. Observe for growth at 2- and 7-day intervals and record your results on the report sheet.

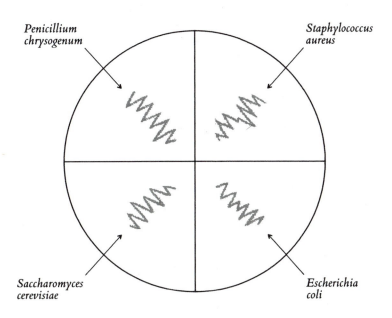

Figure 23-1 Inoculation scheme.

QUESTIONS

1. Between what approximate values on the pH scale do most biological reactions occur?
2. If an organism grows at pH 2, would you expect that the pH of the interior of the cell corresponds to that of the environment? Why? How would you determine this internal pH experimentally?
3. What is the interrelationship of pH and temperature as environmental influences?

Effects of pH of Medium on Growth

Name _____

Desk No. _____

Record results, using a scale from 0 (no growth) to ++++ (very heavy growth). Remember that the growth of the bacteria and the yeast will look quite different from that of the mold.

pH	Staphylococcus aureus *epi*		Escherichia coli		Saccharomyces cerevisiae		Penicillium chrysogenum	
	2 days	7 days	2 days	7 days	2 days	7 days	2 days	7 days
3.0		○		○		○		○
5.0		○		○		+		6
7.0		+++		+++		○		+++
9.0		++		+++		○		++

24. *Effects of Energy Source and Buffer on Growth*

The growth of a microorganism in a culture medium depends partly on the amount of available energy.

As microorganisms attack substrates, various end products accumulate. Some, like those that radically change the pH of the growth medium, can limit further growth. Many of the end products of carbohydrate breakdown are organic acids that lower the pH. In contrast to this, the oxidation of salts of organic acids raises the pH. To prevent inhibitory fluctuations in pH, microbiologists add **buffers** to growth media. Buffers are substances that decrease the changes in pH that result from the production of acids and alkalies. Common buffers include salts of phosphates, carbonates, and various organic materials such as proteins.

In this exercise you will observe the influence of the amount of available glucose on the growth of a species of *Streptococcus*. This coccus uses glucose as its primary energy source and forms lactic acid as an end product. Since the unchecked accumulation of acid lowers the pH and limits further growth, you will use phosphate as a buffer.

PROCEDURE

1. Inoculate the following media with a culture of *Enterococcus faecalis*.
   (a) 1% tryptone + 1% yeast extract
   (b) 1% tryptone + 1% yeast extract + 1% glucose
   (c) 1% tryptone + 1% yeast extract + 1% glucose + 0.5% K_2HPO_4.
   (d) 1% tryptone + 1% yeast extract + 0.1% glucose + 0.5% K_2HPO_4.
2. Incubate for 2 days at 37°C.
3. Tap the tubes to suspend the organisms uniformly and compare the growth (turbidity) by visual inspection or compare the cell crops on a turbidimeter.
4. Use the pH meter or pH indicators provided to determine the pH of each tube (Figures 24-1 and 24-2).
5. Record your results on the report sheet.

Figure 24-1   A pH meter.

Figure 24-2 Testing the pH of a liquid with pH paper.

QUESTIONS

1. What serves as an energy source in medium (a)?

2. How does K_2HPO_4 function as a buffer?
3. What type of reaction would raise the pH of a growth medium?

Effects of Energy Source and Buffer on Growth

Name _____

Desk No. _____

Record results, using a scale from 0 (no growth) to ++++ (very heavy growth).

Medium	Growth	Final pH
(a) 1% tryptone + 1% yeast extract		
(b) 1% tryptone + 1% yeast extract + 1% glucose		
(c) 1% tryptone + 1% yeast extract + 1% glucose + 0.5% K_2HPO_4		
(d) 1% tryptone + 1% yeast extract + 0.1% glucose + 0.5% K_2HPO_4		

MICROBIAL NUTRITION

The nutritional requirements of microorganisms vary widely. **Autotrophic** organisms, including those that are photosynthetic, can synthesize their cell metabolites entirely from inorganic compounds; **heterotrophic** species require one or more organic nutrients.

Some heterotrophic microorganisms can synthesize all the amino acids, vitamins, and other compounds essential to living cells from inorganic starting materials such as nitrogen salts, if they have an organic source of energy and carbon. Other, more fastidious heterotrophs have nutritional requirements as complex as, or more complex, than those of higher organisms. *Streptococcus pyogenes*, for example, the organism that causes scarlet fever and septic sore throat, needs fifteen amino acids, whereas humans appear to require only nine! The vast majority of heterotrophic microorganisms fall between such extremes, but even within this nutritional middle ground, there is wide variation from species to species.

These differences in nutritional requirements reflect differences in synthetic abilities. The ability to use a given compound as an energy source or to use various inorganic materials for synthesizing proteins and other cytoplasmic components depends on the presence of a series of enzymes, without which the cell becomes more nutritionally demanding. The formation of these enzymes is directly controlled by the genetics of the cell. The lack or repression of one or more genes that code for the formation of these enzymes is directly reflected in the nutritional requirements of the cell.

Occasionally (perhaps one cell in every million), in a bacterial population, a cell will change, or **mutate**, in such a way that it loses the ability to synthesize a vitamin, amino acid, or other essential cell component. Such losses of synthetic capacity represent alterations (mutations) in the genetic makeup of the cells. Deficient mutants in a population are not usually apparent because they continue to grow along with the other cells of the population as long as the growth medium provides them with the missing essential nutrient or nutrients. If the growth environment is deficient in the necessary nutrient, the mutation becomes lethal: the variant starves while the normal progeny of the parent strain continue to grow and proliferate. Using special techniques, we can isolate mutants that require the addition of one or more amino acids or vitamins to the growth medium in which the rest of the population is cultured. Mutants that require a nutrient that is not needed by the parent strain are termed **auxotrophs**; cells of the parent strain are called **prototrophs**.

25. Lactobacillus *Bioassay for Niacin*

The basic nutritional requirements of some of the more fastidious heterotrophs are:

1. A source of energy and material for certain cell structures – for example, carbohydrates
2. A source of cell-building materials, enzymes, and nucleic acids – for example, amino acids, purines, and pyrimidines
3. Accessory growth factors for metabolism and regulatory mechanisms – for example, vitamins
4. Mineral salts

The amount of growth of nutritionally fastidious bacteria in a medium, up to a certain maximum, is proportional to the concentration of the essential nutrients. In most media they are supplied in great excess, but a shortage of any one of them limits growth. The microbiological assay of amino acids and vitamins is based on this fact.

For a microbiological assay you use a chemically defined medium that is known to be deficient in the nutrient being assayed. By adding small amounts of that nutrient, you can observe a growth response proportional to the concentration of the nutrient. If you want to examine some material for the presence of a certain vitamin, you add an aliquot of the material to a medium deficient in the vitamin being assayed and then measure the extent of the growth response of the assay organism, which indicates the vitamin content of the food.

In this exercise you will inoculate a bacterial strain (the assay organism) that requires the vitamin niacin into a niacin-deficient medium and into other media that contain various concentrations of niacin. After incubation you will observe that the amount of growth and of acid production is directly related to the concentration of the vitamin.

PROCEDURE

You are provided with a niacin-requiring strain of *Lactobacillus plantarum* that has been grown in a medium containing niacin. (After incubation, the culture was centrifuged and washed to remove any niacin that might have been adhering to the cells.) You are also provided with eight 10-ml tubes of assay medium with the following concentrations of niacin: 0.0, 0.025, 0.05, 0.1, 0.15, 0.2, 0.3, and 0.5 micrograms (μg) of niacin per tube.

1. Label the tubes 1 through 8. Holding a 1.0 ml pipette vertically, add one drop of *Lactobacillus plantarum* cells to each tube.
2. Incubate at 37°C for 24 hours.
3. Using an uninoculated tube of niacin assay medium as a standard, set the spectrophotometer at 100% transmittance, using a wavelength of 650 nanometers.
4. Read the inoculated tubes in the spectrophotometer and then plot absorbance versus niacin concentration.
5. Reincubate these tubes at 37°C for another 24 hours.
6. Add 0.1 ml of bromthymol blue indicator to each tube. Titrate each tube with 0.1 N NaOH, shaking the tube constantly until pH 7 (green-blue color) is reached. Add 0.1 ml more of indicator for each 1 ml of NaOH added. If the titration volume exceeds the

Figure 25-1   Titration to measure the quantity of acid.

capacity of the tube, use flasks to continue the titration. Return the titrated sample to the tube for color matching. Bring all tubes to the same color (Figure 25-1).

7. On the report sheet, record the amount of NaOH used for each titration and plot the concentration of niacin versus ml of NaOH.

QUESTIONS

1. What could be a possible explanation for growth in tube 1?
2. Why is an uninoculated tube used as the standard in turbidity readings?
3. Is there linearity at all times between absorbance and concentration of niacin? Why?

Lactobacillus Bioassay for Niacin

Name _____

Desk No. _____

Record the absorbances and milliliters of NaOH used to titrate media at each concentration of niacin.

µg of niacin per tube	Absorbance	ml of 0.1 N NaOH
0.0		
0.025		
0.05		
0.1		
0.15		
0.2		
0.3		
0.5		

Absorbance

Niacin

0.1 N NaOH (milliliters)

26. Neurospora *Bioassay for an Amino Acid*

In this experiment we touch on one of the historical roots of molecular biology. In 1941 George Beadle and Edward Tatum isolated a number of genetic mutants of the mold *Neurospora*. This achievement opened the way to the analysis of mutations at the molecular level. They shared the Nobel Prize in 1958 in recognition of this work.

Genetic mutations that cause a microorganism to lose a biosynthetic ability have been of great value in the study of microbial metabolism and physiology. As a result of the loss of ability to synthesize one particular enzyme, the mutant organism then requires an external supply of the end product of the reaction catalyzed by that enzyme. Auxotrophic strains respond not only to the end products of the missing reactions but also to those compounds that are farther along in the biosynthetic sequence of reactions that synthesize the essential metabolite. The substitution of intermediate and end products is widely used in delineating biosynthetic reaction sequences and interlocking metabolic schemes.

In the following experiment, you will study the growth response of an auxotrophic mutant of the mold *Neurospora crassa* (ATCC 10780). This mutant has a genetic block in the biosynthetic sequence for methionine:

$$\text{cysteine} \rightarrow \text{cystathionine} \rightarrow \text{homocysteine} \rightarrow \text{methionine}$$

You will measure the amount of growth of the organism in the presence of increasing concentrations of methionine and in the presence of two precursors in the biosynthetic sequence: cysteine and homocysteine. The site of the genetic block can be determined from these data. It is important to test these two precursors because a bioassay is valid only if growth is directly proportional to the presence of a specific metabolite.

PROCEDURE

1. You will be given *Neurospora* basal medium in microfernbach flasks supplemented with:
   (a) No supplement
   (b) Methionine, 5.0 μg/ml
   (c) Methionine, 10 μg/ml
   (d) Methionine, 20 μg/ml
   (e) Methionine, 40 μg/ml
   (f) Methionine, 50 μg/ml
   (g) Cysteine, 50 μg/ml
   (h) Homocysteine, 50 μg/ml

2. With a sterile inoculating loop, remove some of the spores from a *N. crassa* culture and suspend them in 5 ml of sterile saline solution. Mix this suspension thoroughly; it should be just visibly turbid to the naked eye.

3. Inoculate the flasks with 0.1 ml of the spore suspension. Incubate the flasks at 30°C in a dark incubator for 72–96 hours.

4. After incubation, add 3 ml of formaldehyde solution (40%) to the first flask, shake it, and pour its contents through a funnel filled with clean filter paper.

5. When all the culture medium has drained off, remove the filter paper from the funnel, roll the mycelium into a compact ball, and express the water against a dry piece of filter paper. Place the damp, flattened mycelial pad into an appropriately labeled depression in a spot plate.

6. Repeat this process for all flasks. Dry the pads for 60 minutes at 60°C or under a heat lamp. Weigh the dried mycelial pads and record their weights on the report sheet. Plot the weights on the graph.

QUESTIONS

1. What is the locus of the genetic block in methionine synthesis in this mutant? What is the proof?
2. What are the requirements for a bioassay to be a valid test?
3. What other methods could be used to measure the growth of *Neurospora*?
4. Which methods for measuring microbial growth would probably be invalid for measuring the growth of *Neurospora*? Why?

Neurospora Bioassay for an Amino Acid

Name _____

Desk No. _____

Flasks	Weight of mycelium
(a) No supplement	
(b) Methionine 5 μg/ml	
(c) 10 μg/ml	
(d) 20 μg/ml	
(e) 40 μg/ml	
(f) 50 μg/ml	
(g) Cysteine 50 μg/ml	
(h) Homocysteine 50 μg/ml	

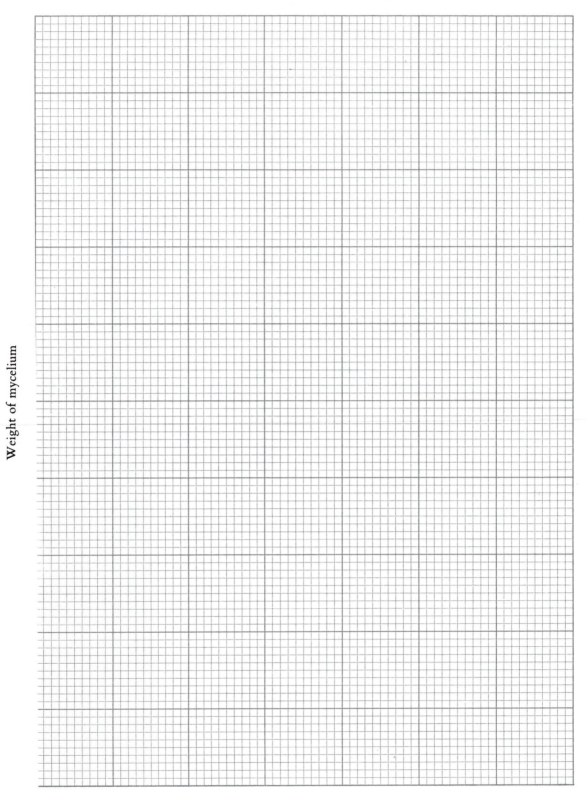

Weight of mycelium

Concentration of methionine

27. *Bacterial Photosynthesis*

Photosynthetic bacteria, like algae and plants, utilize light energy for synthesis and growth. The number and types of photosites and photosynthetic pigments and the nature of the hydrogen donor for the photoreduction of CO_2 vary among photosynthetic bacteria.

For example, in the blue-green bacteria, phycobiliproteins, chlorophylls, and carotenoids are used in the photosynthetic process, whereas the pigments of the purple and green photosynthetic bacteria consist of bacteriochlorophylls and carotenoids. (The characteristic colors of these bacteria indicate the relative concentrations of these pigments; the carotenoids generally predominate over the bacteriochlorophylls.) The photosynthetic pigment of the halobacteria is bacteriorhodopsin (so named because of its similarity to the rhodopsin of the retina).

Photosynthesis by the blue-green bacteria is an aerobic process that involves the photolysis of water and evolution of oxygen, but other types of bacterial photosynthesis are anaerobic. The blue-green bacteria, like plants and algae, use water as the source of reducing power, whereas other bacterial

Figure 27-1 Electron micrograph of the fine structure of a photosynthetic bacterium, *Rhodospirillum rubrum*. The electron-translucent bodies are the chromatophores, the sites of photosynthetic pigments. (From A. E. Vatter and R. S. Wolfe, *J. Bact.* 75:484, 1958.)

photosynthetic processes use substrates including H_2, sulfides, and organic compounds as reductants.

The hydrogen donor (reductant) for the purple and green sulfur bacteria is one of several sulfide compounds (e.g., H_2S). The purple nonsulfur bacteria and the halobacteria use organic compounds (e.g., isopropanol) as both the reductant and the carbon source for the photosynthetic process. Furthermore, in the absence of light, the halobacteria are not obligately photosynthetic since in aerobic conditions they are chemoheterotrophic (chemorganotrophs) and use organic compounds as sources of energy for cell synthesis.

In this exercise you will observe the photoheterotrophic and chemoheterotrophic growth of the purple nonsulfur bacterium *Rhodopseudomonas.* (A photomicrograph of another photosynthetic bacterium, *Rhodospirillum rubrum,* is reproduced in Figure 27-1.)

PROCEDURE

1. Aseptically fill two 250-ml sterile flasks to a depth of about 1 centimeter with Lascelles' medium. The media in these two flasks will be used as shallow layers for aerobic growth. For anaerobic growth, fill two 250-ml flasks with Lascelles' medium to the base of the flask's neck.
2. Inoculate each of the four flasks with 0.1 ml of *Rhodopseudomonas spheroides.*
3. Cover two of the flasks (one for aerobic growth and one for anaerobic growth) with aluminum foil or a heavy black plastic bag to keep out light.
4. Incubate all four flasks at 30°C under a light source for 2–7 days until growth appears.
5. Remove the foil and compare the growth and pigmentation in the four flasks.
6. Record your results on the report sheet.

QUESTIONS

1. Is the amount of pigmentation the only change that shows growth under the varying conditions?
2. What is meant by cyclic photophosphorylation?
3. Where and under what conditions do you find the photosynthetic halobacteria?

Bacterial Photosynthesis

Name _____

Desk No. _____

Record results, using a scale from 0 (no growth, no pigment production) to $++++$ (very heavy growth, very heavy pigment production).

Flask	Growth	Pigment production
Aerobic in light		
Aerobic in dark		
Anaerobic in light		
Anaerobic in dark		

28. *Relation of Free Oxygen to Microbial Growth*

Bacteria vary considerably in their requirements for gaseous oxygen (Figure 28-1). Some bacteria do not grow in the absence of oxygen; others do not develop in its presence; some organisms can adapt to either the presence or absence of oxygen. An organism that requires oxygen is called **aerobic**; one whose growth is inhibited by oxygen is **anaerobic**; one that can grow under both aerobic and anaerobic conditions is termed a **facultative anaerobe**.

Few microbes are strictly aerobic or anaerobic; most are somewhere between. Microorganisms show a constant gradation with respect to oxygen requirements. Many terms other than aerobic, anaerobic, or facultative are used to describe the oxygen requirements of the various species. For example, **microaerophilic** organisms require free oxygen but at a very low concentration. **Aerotolerant** bacteria are anaerobic bacteria that grow at oxygen pressures less than atmospheric pressure.

In addition to serving as a final acceptor of electrons in cell respiration, oxygen also alters the oxidation–reduction potential of cells. Many enzyme systems in bacteria require strongly reduced conditions, that is, a low oxidation–reduction potential in order to function. Others require oxidized conditions, a high oxidation–reduction potential.

In this experiment you will use agar-shake cultures. With this method you observe the *location* of the area of growth in the tube as the index of the oxygen requirements of microorganisms.

YEAST-EXTRACT – TRYPTONE AGAR (YTA) SHAKES

PROCEDURE

1. Melt three tubes of yeast-extract–tryptone agar (YTA) and hold at 100°C for 10 minutes to expel dissolved oxygen.
2. Cool to a temperature of 42–45°C and inoculate each heavily (several loopfuls) with one of the cultures of *Escherichia coli*, *Micrococcus luteus*, and *Clostridium perfringens*. Gently whirl the inoculated agar to distribute the inoculum, being careful not to agitate the tubes in a way that would incorporate air in the agar.
3. Solidify the agar by placing the tubes in cool water (below 40°C).
4. Incubate at 37°C for 2 days and observe the location and appearance of growth (Figure 28-2).
5. Illustrate and label your results on the report sheet.

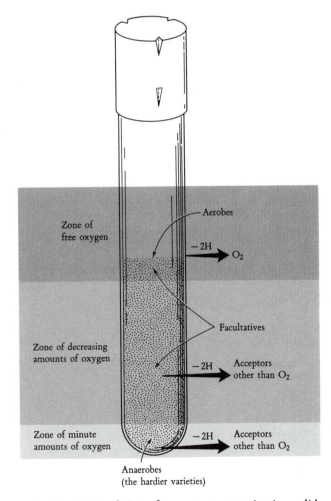

Figure 28-1 Relation of oxygen concentration in a solid medium to types of microbial growth.

Figure 28-2 Locations of microbial growth in agar-shake cultures.

BROMCRESOL-PURPLE (BCP)– GLUCOSE SHAKES

Bromcresol purple (BCP) is a pH indicator that is purple at pH higher than 6.8 and yellow at pH lower than 6.8. In this experiment you use a glucose agar medium, to which BCP has been added, to determine metabolic characteristics and O_2 relationships of certain bacteria.

PROCEDURE

1. Melt 3 tubes of BCP–glucose agar. (*Note*: Other carbohydrates may be substituted if desired.) Temper to 45°C. An uninoculated tube will serve as a control.
2. Inoculate each tube heavily with one of the

Table 28-1 Typical Reactions in BCP-Glucose Shakes

Observation	Conclusion
Growth on top of agar only	Organism is an obligate aerobe; requires O_2
Growth throughout tube	Organism is a facultative anaerobe; can grow with or without O_2
Growth only in bottom or lower ¾ of tube	Organism is an obligate anaerobe; O_2 is toxic to it
Color of tube: No change, purple throughout	No growth, or organism oxidizes glucose to CO_2 and H_2O, no acid
Yellow, and growth only at top	Acid is produced from glucose oxidatively
Yellow, growth in anaerobic portions of shake	Glucose is fermented with acid production
Growth throughout, yellow at bottom but purple on top	Organism ferments glucose and the acid produced is oxidized to CO_2 and H_2O is present; this is called "alkaline reversion"
There are cracks in the agar.	Gas is produced by the fermentation of glucose.

cultures of *E. coli, Micrococcus luteus,* and *Clostridium perfringens.*

3. Mix each tube gently after inoculation by rolling the tube between your palms. *Do not incorporate bubbles while mixing.*

4. Allow tubes to solidify at room temperature. Incubate as instructed.

5. After incubation, observe your tubes for the reactions shown in Table 28-1. Record observations and conclusions on the report sheet. Be sure to use the uninoculated tube as a control when observing your tubes.

QUESTIONS

1. What is oxidation? Biological oxidation?
2. What is respiration? Fermentation? Anaerobic respiration?

Relation of Free Oxygen to Microbial Growth

Name _____

Desk No. _____

Illustrate the location and appearance of growth. On the lines above the drawings, write the names of the organisms and whether they are facultative, anaerobic, or aerobic.

YTA SHAKES

BCP – GLUCOSE SHAKES

Microlutlus
Faculative

P fluro
Faculative

E coli
anaerobic

29. *Anaerobic Culture of Bacteria*

The effect of atmospheric oxygen on microbial growth is closely related to the oxidation–reduction potential of the culture medium. The **O–R potential** is a means of expressing the degree of oxidation or reduction of a compound or an environment. A compound with a high ratio of H to O (higher than H_2O) is strongly reduced and has a **negative O–R** potential. A highly oxidized material has a **positive O–R** potential. Anaerobic bacteria can be grown by eliminating free oxygen from the environment or by establishing a low O–R potential by providing sufficient reducing materials in the growth medium. This exercise will illustrate the relationships of O–R potential to growth.

A considerable range of oxygen tolerance exists among the "strict" anaerobes, some growing in air to a barely visible extent and others requiring the complete exclusion of free oxygen.

The protection of anaerobic cultures from free oxygen can be accomplished in several ways.

1. You can expel oxygen gas from culture media by boiling and then prevent its reintroduction by adding a relatively solid barrier such as a layer of vaspar or mineral oil. This is useful with media such as those used for fermentation tests. Boil the media for 10 minutes, cool rapidly to 40–45°C, and inoculate heavily (a drop or several loopfuls) without agitation. Put the seal on immediately after inoculation.
2. You can use various media that contain a reducing agent such as sodium thioglycollate ($HSCH_2COONa$), which reacts with oxygen and keeps the enzymes in the cell in a reduced condition. It is used in broth media (which need not be sealed) and in agar for plating.
3. You can remove oxygen from the atmosphere of a closed container by several means.
   (a) One method uses a Brewer anaerobic jar or some similar device. The plates or tubes to be incubated are put in the jar with a commercial preparation that releases hydrogen and carbon dioxide when exposed to water (Figure 29-1).

Figure 29-1 Brewer anaerobic jar, using hydrogen gas pack. (From J. H. Brewer and D. Allgeier, *Appl. Microbiol.* 14:986, 1966.)

Screw clamp
Gasket
Rubber boot
Pellet catalyst
Catalyst holder
Hydrogen gas pack
Culture plates
Anaerobic indicator

Catalyzed by platinum, the hydrogen and any oxygen present in the jar react to form water, which releases more hydrogen.
   (b) Replacement of air by an inert gas is a widely used method and offers a number of advantages. Agar plates or tubes to be incubated are placed in an anaerobic incubator. The incubator is evacuated, and carbon dioxide, hydrogen, or nitrogen is run in (Figure 29-2).
4. When cultures and other materials must be manipulated in the absence of oxygen, a **glove box** (Figure 29-3) is used. This is a sealed unit in which air can be replaced by a particular gas mixture (e.g., CO_2/N_2), which provides the proper atmosphere for the survival and growth of the microorganisms being manipulated. Work inside is done by using sealed-in rubber gloves. The

Figure 29-2 Anaerobic incubator used to provide various gas atmospheres.

box can be fitted with internal fixtures to provide gas, ultraviolet light, and electricity.

5. More specialized techniques are necessary for isolating and culturing oxygen-sensitive anerobes.* These involve prereduced, anaerobically sterilized media that are maintained in tightly sealed anaerobic culture tubes or bottles. To ensure that the oxygen content of the medium remains depleted, a reducing agent such as cysteine is added. The medium also contains an oxidation–reduction dye such as resazurin as an indicator to show that the medium is anaerobic. Because of the fastidious nature of many anaerobes, all manipulations in this method of culture are carried out under a stream of oxygen-free gas to maintain the anaerobic atmosphere.

Inoculations are made into the agar media while it is a liquid at 45°C by either a closed or an open method. In the closed method (Hungate technique), the inoculum is introduced by a syringe and needle through a rubber stopper that seals the tube. During inoculation by the open method, the rubber stopper is removed and the open tube continuously flushed with oxygen-free gas while the medium is inoculated with a loop

---

*These techniques are often referred to as the Hungate technique (after Dr. R. E. Hungate, who pioneered this method) and the VPI method (after Dr. W. E. C. Moore and co-workers at Virginia Polytechnic Institute where these procedures were modified and developed).

Figure 29-3 Plastic bag glove box, with automatic microcomputer air-lock controller, which is used in handling anaerobic cultures. (Courtesy of Coy Labs.)

or Pasteur pipet. After inoculation, the tube is immediately sealed to retain the anaerobic environment. Each tube is then rolled in a tube-rolling device until the agar medium solidifies on the walls of the tube. Following incubation, isolated colonies develop on the agar layer (Figure 29-4).

The organism in this experiment is a species of the genus *Clostridium*, used because it forms endospores, which are more resistant to adverse environmental conditions than vegetative cells. Endospores do not grow in the presence of O_2 though they can remain viable and germinate when their environment becomes conducive to growth.

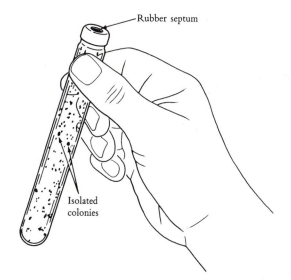

Figure 29-4 Balch tube showing isolated colonies.

PROCEDURE

1. You are provided with one melted glucose –BCP agar shake from the waterbath, one thioglycollate (thio) broth, 2 thio plates, two PCA plates, and one SB broth.
2. Label all media tubes and plates.
3. The *Clostridium* is provided in bottles to be shared by several students. Remove approximately 1 ml of culture from the bottle with a sterile pipet. *Be very careful* to remove cells from the bottom of the bottle and not to bubble O_2 into the culture through the pipet.
4. Inoculate all labeled plates, broths, and shakes with 2 or 3 drops of *Clostridium* culture. Discard any leftover culture and put your pipet in the pipet discard jar. Do not put it back into the inoculum bottle!
5. Using your inoculating loop, spread the inoculum across the plates. Isolation is not important, so don't worry about your aseptic technique.

6. *Very gently* mix the shake and broths: You do *not* want to incorporate O_2 into the media.
7. Incubate one thio plate, one PCA plate, the thio broth, SB broth, and BCP shake in your incubator box at 30°C. Place the other thio plate and PCA plate in the Brewer anaerobic jar for your instructor to incubate.

QUESTIONS

1. Why does the oxygen that dissolves in thioglycollate broth fail to inhibit the growth of the anaerobes?
2. Does the *Clostridium* grow the same way in the different growth conditions?

Anaerobic Culture of Bacteria

Indicate growth, plus any other activity such as gas production.

Medium	Clostridium sp.
Glucose BCP agar shake	
Thioglycollate broth	
SB broth	
Thioglycollate plate in air	
Thioglycollate plate in Brewer jar	
PCA plate in air	
PCA plate in Brewer jar	

30. *Antiseptic and Disinfectant Action*

In our struggle to control detrimental microbes, we have discovered thousands of inorganic and organic chemicals that are toxic to microorganisms. These chemical agents either inhibit microbial activities and growth or kill the microorganisms. Inhibitory chemicals, usually used to cleanse the skin, are termed **antiseptics**. Bacteriostatic agents inhibit growth, and **disinfectants** are bacteriocidal. These agents can be alcohols, halogens, aldehydes, phenols, heavy metal complexes, or quaternary ammonium compounds. Alcohols coagulate essential proteins and chlorine oxidizes them; formaldehyde combines with amino groups, and heavy metals such as Hg and Ag inhibit enzymes. Phenolics and quarternary ammonium compounds disrupt cytoplasmic membranes. Many of these agents are ineffective in the presence of high concentrations of organic matter, soaps, or detergents. See Figure 30-1 for the structure of the widely used phenolic compound, hexachlorophene, and a quarternary ammonium compound, benzalkonium chloride. This exercise shows you how to compare the antiseptic and disinfectant actions of several toxins.

PROCEDURE

1. Obtain two tubes of melted nutrient agar that have been cooled to 45°C in a water bath.

2. Inoculate one tube with 0.5 ml *Escherichia coli* and the other with 0.5 ml *Bacillus subtilis*. Mix well and pour into sterile petri plates. Allow the agar to solidify.

3. With alcohol-flamed forceps distribute six filter-paper discs around each seeded petri plate (Figure 30-2).

4. Using the pipettes provided in the antiseptic and disinfectant solutions, add 0.1 ml of the various agents to the discs on each plate and label the bottom of the plate with the identity of each agent. If you brought your own agents to the laboratory, test them in the same way.

5. Incubate at 37°C until the following laboratory period.

6. Observe each plate and measure the diameters of the areas of inhibition of growth in millimeters (Figure 30-3). Enter the results in the table on your report sheet.

This method of comparison can be useful but is imprecise. To obtain an accurate measure of inhibitory and lethal efficiency, you would have to consider such factors as the bacterial concentration, disinfectant concentration, temperature, pH, and relative water solubility of the agents tested. Furthermore, various chemicals diffuse at different rates and are inactivated to varying degrees by culture media. Some materials, such as the quaternary ammonium compounds, cannot be tested at all by this method because they react with the medium, producing ambiguous results.

For the testing of the effectiveness of a disinfectant, the standard is phenol, and the relative effectiveness of a new chemical is compared with that of phenol under standardized conditions. The ratio of the toxicity of the test disinfectant to that of phenol for the standard test organisms, *Salmonella typhi* (gram −) and *Staphylococcus aureus* (gram +), is called its **phenol coefficient** and is an expression of its effectiveness.

The causes of the toxicity of most antiseptics and disinfectants are poorly understood. Each, however, adversely affects the cell either by disrupting its physical makeup or by blocking its energy-yielding or synthetic processes.

Hexachlorophene
(a bis-phenol)

Benzalkonium chloride
(a quarternary ammonium disinfectant)

*n ranges from 8 to 18

Figure 30-1 Disinfectants.

(1) Place discs on seeded agar   (2) Add antiseptics and disinfectants

Figure 30-2   Method of applying antiseptic or disinfectant to a plate seeded with bacteria.

Figure 30-3   Zones of inhibition around antiseptics and disinfectants.

QUESTIONS

1. This exercise does not distinguish between inhibition and killing. Design an experiment that would enable you to distinguish between the two.
2. Were any microorganisms growing within the zone of inhibition of any of the antiseptics or disinfectants? Explain why.
3. What factors, other than the chemical, affect the size of the zone or inhibition?
4. List three possible modes of action of these inhibitory compounds.
5. List three chemical classes to which these inhibitory compounds belong.

Antiseptic and Disinfectant Action

Record the diameter of the zone of inhibition for each compound tested.

Antiseptic or disinfectant	Concentration	Diameter of inhibition zone (mm)	
		Escherichia coli	Bacillus subtilis
1. phenol	5 %	42 mm	none
2. Clorox		clear	50 mm
3. mecurochome		27 mm	31 mm
4. 70% isopropanol		37 mm	none
5.			
6.			

31. *Lethal Action of Ultraviolet Light; Photoreactivation*

Although certain wavelengths of visible light are beneficial and even essential to some bacteria, such as the photosynthetic forms, sunlight is usually injurious to most bacteria. This effect is due primarily to the ultraviolet (UV) portion of the spectrum, in particular to light of wavelengths between 240 and 300 nanometers (nm); the peak of toxicity for most microorganisms is near 265 nm.

Deoxyribonucleic acid has a maximum absorption peak in the UV range at 265 nm, and radiation at this wavelength markedly affects mutation rates. UV radiation causes **dimer** formation, chemical bonding between adjacent pyrimidine bases on one DNA strand or (sometimes) between DNA strands, which interferes with the replication and function of the DNA molecule. The irradiated bacterial cell thus cannot reproduce and dies.

The bactericidal effect of exposure of a culture to ultraviolet light can be reduced by immediate exposure to visible light of wavelengths 365–450 nm (Figure 31-1). This reversal of killing action of ultraviolet light is called *photoreactivation*. In photoreactivation, visible light activates an enzyme that cleaves the pyrimidine dimers that result from UV irradiation, thereby enabling the DNA to function. Thus, the number of surviving cells in a population that is exposed to visible light immediately after exposing to UV light is many times that of a population exposed solely to UV light. However, there are always some cells that cannot be repaired by photoreactivation.

In this exercise you will observe irradiated bacterial cultures for the lethal action of UV light and photoreactivation by visible light.

PROCEDURE

Steps 1–6 in this procedure and in Figure 31-2 correspond.

1. Make pour plates of 10^{-6}, and 10^{-7}, and 10^{-8} dilutions of a 24-hour culture of *Escherichia coli* on trypticase soy agar. From the colony counts of these plates at the next laboratory period, you will determine the average number of cells per milliliter in this culture.
2. Place 2 ml from the 10^{-6} dilution blank into each of four small, sterile, petri plates. Label these plates 5 sec, 10 sec, 15 sec, and 20 sec.
3. Place the plates in order of their increasing times of exposure in a row on the white taped line in the UV box. Remove all the petri dish lids and turn on the UV light.

Figure 31-1 Light spectrum.

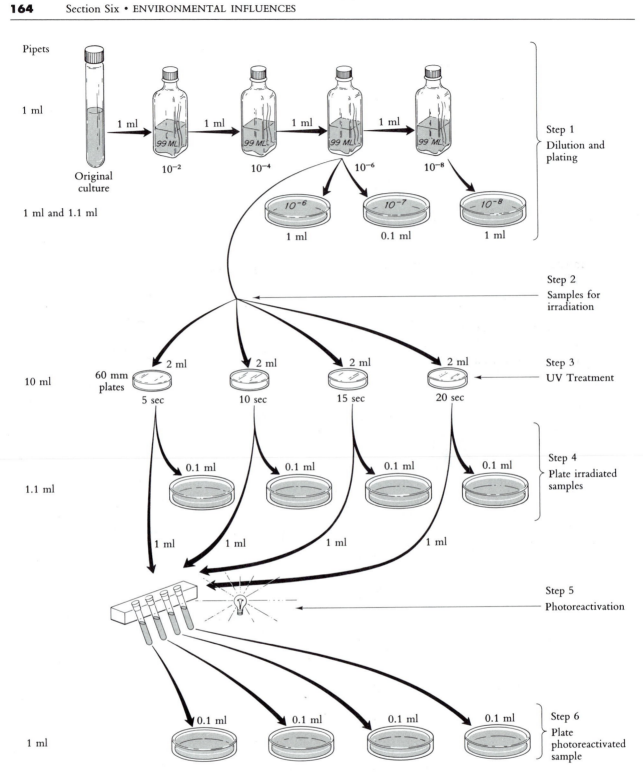

Pipets

1 ml

1 ml and 1.1 ml

Original
culture

1 ml

1 ml

1 ml

10^{-2}

10^{-4}

10^{-6}

10^{-8}

10^{-6}

10^{-7}

10^{-8}

1 ml

0.1 ml

1 ml

Step 1
Dilution and
plating

Step 2
Samples for
irradiation

10 ml

60 mm
plates

2 ml

2 ml

2 ml

2 ml

Step 3
UV Treatment

5 sec

10 sec

15 sec

20 sec

1.1 ml

0.1 ml

0.1 ml

0.1 ml

0.1 ml

Step 4
Plate irradiated
samples

1 ml

1 ml

1 ml

1 ml

Step 5
Photoreactivation

1 ml

0.1 ml

0.1 ml

0.1 ml

0.1 ml

Step 6
Plate
photoreactivated
sample

Figure 31-2 UV – photoreactivation dilution scheme.

Replace the lids and remove the plates at the appropriate 5-second intervals. Wrap with aluminum foil to prevent possible photoreactivation (Figure 31-3).

4. Immediately after its exposure to the UV light, prepare a 10^{-7} dilution of each sample by transferring a 0.1 ml aliquot from the exposed plate to a sterile petri plate, adding the proper amount of melted and tempered trypticase soy agar and mixing by swirling. Label these new plates with the word *irradiated*, the time of exposure, and the dilution (10^{-7}). Wrap in aluminum foil.

5. Transfer a 1.0 ml sample from each plate irradiated in Step 3 to a different sterile test tube; label each with the time of irradiation.

6. Place the four tubes 6 inches away from a 500-watt bulb for 30 minutes (Figure 31-4). If the light source you are given does not have a fan attached for cooling, place the tubes in a beaker of ice.

7. Plate 10^{-7} dilutions of each sample exposed to the light, as outlined in Step 4. Label each plate with the word *photoreactivated*, the time of UV irradiation, and the dilution 10^{-7}.

8. Incubate all plates at 30°C for 48 hours.

9. After incubation, count and record the number of colonies on each plate. Calculate the percentage of survivors in the irradiated and photoreactivated samples.

10. On your report sheet, plot the number of survivors versus time for the irradiated and photoreactivated samples.

Figure 31-3 Exposing petri plates to ultraviolet light.

Figure 31-4 Photoreactivating irradiated cells.

QUESTIONS

1. What other types of radiation are lethal to microbes?
2. Is the photoreactivation phenomenon limited to microbes?
3. Why were the survivor curves plotted on semilog paper?
4. What is "dark repair"?
5. Can you determine whether the microorganisms were killed as the result of single direct hits or multiple hits?

Lethal Action of Ultraviolet Light; Photoreactivation

Name _____

Desk No. _____

Record results.

In the original culture there were _____ $\times 10^7$ organisms per milliliter.

Time of irradiation	Irradiated cells		Photoreactive cells	
	No. of surviving organisms $\times 10^7$	Percent survivors	No. of surviving organisms $\times 10^7$	Percent survivors
5 seconds				
10 seconds				
15 seconds				
20 seconds				

Use the semilog graph on the reverse side to plot the number of irradiated survivors per milliliter versus time and the number of photoreactivated survivors per milliliter versus time.

30 sec

1 min

2 min

3 min   6 colonies

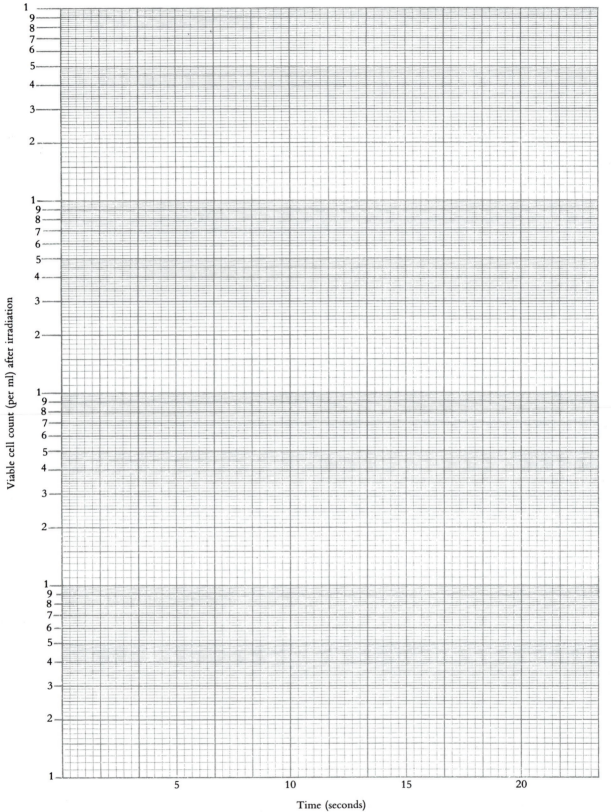

Viable cell count (per ml) after irradiation

Viable cell count (per ml) after photoreactivation

Time (seconds)

32. *Antimetabolites: Inhibition by Sulfanilamide*

Many chemicals are bacteriostatic or bactericidal through inhibition of the bacterial cell's growth, that is, interference with the manufacture or utilization of some essential chemical, or **metabolite**. If a chemical has a structure that is analogous to that of an essential metabolite of the cell (an analogue), it can replace the metabolite in one or more of the enzymatic reactions and thus block its normal function. Such inhibitory chemicals are called **antimetabolites**.

Among the first antimetabolites recognized as such were the sulfa drugs, of which sulfanilamide is an example. The structure of the sulfa drugs is closely analogous to that of *p*-aminobenzoic acid, which is a metabolite essential to the formation of the vitamin folic acid. Sulfanilamide inhibits the enzymatic incorporation of *p*-aminobenzoic acid into folic acid. The addition of sufficient *p*-aminobenzoic acid to a culture that has been treated with sulfanilamide will compensate for the sulfanilamide inhibition (Figure 32-1).

The addition of folic acid itself reverses the inhibitive effects of sulfanilamide only in some bacterial species; the cells of other species are impermeable to the vitamin. However, the addition of several metabolites whose syntheses require the folic acid coenzyme does reverse the inhibition.

In this exercise you will observe the inhibition of bacterial growth by sulfanilamide and the reversal of this inhibition by some products of those synthetic reactions that require the folic acid coenzyme.

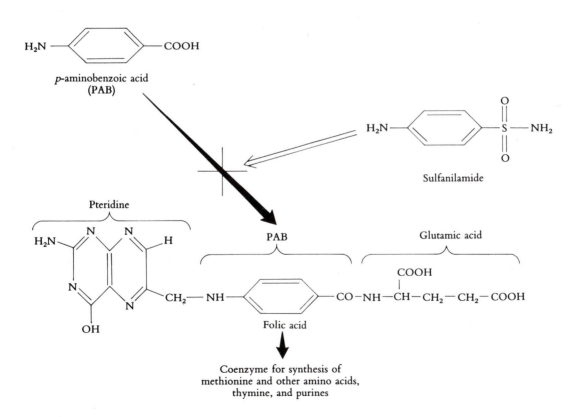

Figure 32-1 Action of sulfanilamide.

PROCEDURE

1. Inoculate the following media with 0.1 of the washed *Escherichia coli* culture provided.
   (a) Minimal broth alone.
   (b) Minimal broth + 0.02 M (final concentration) sulfanilamide.
   (c) Minimal broth + 0.01 M (final concentration) sulfanilamide + 2×10^{-8} M (final concentration) p-aminobenzoic acid.
   (d) Minimal broth + 0.02 M sulfanilamide + 2×10^{-7} M (final concentration) p-aminobenzoic acid.
   (e) Minimal broth + 0.02 M sulfanilamide + 7×10^{-5} M (final concentration) folic acid.
   (f) Minimal broth + 0.22 M sulfanilamide + the metabolites — L-methionine (3×10^{-3} M), thymine (2×10^{-4} M), serine (1×10^{-4} M), and xanthine (7×10^{-5} M).
2. Incubate for 2 days at 37°C.
3. Rotate the tubes to suspend the organisms uniformly and compare the growth visually or turbidimetrically.

QUESTIONS

1. What is meant by competitive inhibition?
2. Give some illustrations of other vitamins or growth factors and their antimetabolites.
3. Why are humans not adversely affected by the moderate use of sulfa drugs in the treatment of disease?

Antimetabolites: Inhibition
by Sulfanilamide

Record results, using a scale from 0 (no growth) to ++++ (very heavy growth).

Medium	Growth
A. Minimal broth	
B. Minimal broth +0.02 M sulfanilamide	
C. Minimal broth +0.02 M sulfanilamide +2 × 10^{-8} M p-aminobenzoic acid	
D. Minimal broth +0.02 M sulfanilamide +2 × 10^{-7} M p-aminobenzoic acid	
E. Minimal broth +0.02 M sulfanilamide +7 × 10^{-5} M folic acid	
F. Minimal broth +0.02 M sulfanilamide +the metabolite mixture	

Interrelations of Microbes

Most microbiological studies in the laboratory are of the properties of pure cultures and of changes in single species. In nature, however, isolated species of microbes rarely occur unless there are unusual environmental conditions (e.g., "red tides"). The survival and activities of a species usually depend on the activities of innumerable other species: beneficial, competitive, and antagonistic.

Beneficial relationships or **mutualism** include **symbiosis**, in which two or more species are mutually dependent; **syntrophism**, in which there are mutual exchanges of benefits between two species, but the relationship is not obligatory; and **commensalism**, in which one species benefits and the other is unaffected.

Antagonistic relationships include **antibiosis**, in which one species produces substances toxic to one or more other species; **parasitism**, in which one species lives and feeds on a host species; **pathogenicity**, in which a parasite causes injury to its host; and **predation**, in which members of one species kill members of another species.

The study of these interrelationships of microbes and of the environments in which they take place constitutes microbial ecology, an almost unexplored field.

The following exercises illustrate some of these interrelationships among microbial populations and among other living forms.

33. *Antibiosis*

Microbial species in a common environment often inhibit each other's growth. Besides excreting end products of carbohydrate metabolism (such as alcohols and acids) that can limit growth, some microorganisms produce more complex and quite different chemicals: **antibiotics**.

ANTIBIOTIC SUSCEPTIBILITY

Since the introduction of penicillin, the first practical antibiotic, streptomycin, chloramphenicol, oxy-tetracycline or chlortetracycline, aureomycin, and many more have been discovered and commercially produced. The use of these antibiotic substances to check the development of pathogenic forms in the body has resulted in an immense annual production of antibiotics by pharmaceutical houses. Basic to the medicinal success of an antibiotic is one requirement: It must be inhibitory to the growth of the pathogenic microorganism but relatively nontoxic to the host.

In this exercise we shall demonstrate penicillin production and inhibitory activity in vitro.

173

PROCEDURE

1. Pour a petri plate of the skim milk agar provided. After the medium has solidified, inoculate it in the center with a loopful of a spore suspension of *Penicillium chrysogenum* (Figure 33-1). Incubate (inverted) at room temperature until the following laboratory period. After growth appears, begin Step 2.

2. Melt a tube of nutrient agar, cool to 45°C, seed it heavily with *Staphylococcus aureus*, and pour it into a sterile plate. When the agar has solidified, sterilize a cork borer in an alcohol flame and cut discs of the *Penicillium chrysogenum* agar from the edge of the colony, from about 1 cm away from the edge of the colony, and from the edge of the plate. Place the discs on the *S. aureus* agar. If there are yellowish droplets on the mold mycelium itself, place one loopful of this liquid on the agar also. Place a sensitivity disc containing two units of penicillin on another area of the plate. Incubate (not inverted) at room temperature. After incubation, observe zones of inhibition and illustrate them on your report sheet.

ISOLATING AN ANTIBIOTIC PRODUCER

That fact that growth of bacterial species can be inhibited by the presence of another microorganism has been recognized since the latter part of the nineteenth century. Many of the antibiotics of clinical importance have been isolated from the ubiquitous soil-dwelling bacteria of the order Actinomycetales. The ecological significance of this phenomenon has been discussed in several essays by the Nobel laureate S. Waksman. In this experiment you will isolate and identify some of these microorganisms.

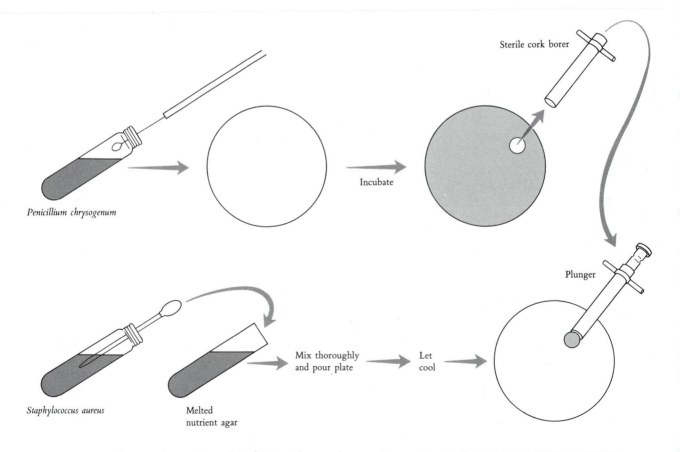

Penicillium chrysogenum

Staphylococcus aureus   Melted nutrient agar

Incubate

Sterile cork borer

Plunger

Mix thoroughly and pour plate

Let cool

Figure 33-1   Procedure for inoculating plates to demonstrate antibiotic susceptibility.

PROCEDURE

1. Mix 1 gram of soil with sterile tap water in a small bottle. Shake vigorously for about 5 minutes (Figure 33-2).

2. Take 1 ml of the sample and dilute in 9 ml dilution blanks to a 1:100 and a 1:1000 dilution. Streak two Emerson agar plates with the dilutions.

3. Incubate the plates for approximately 3 days

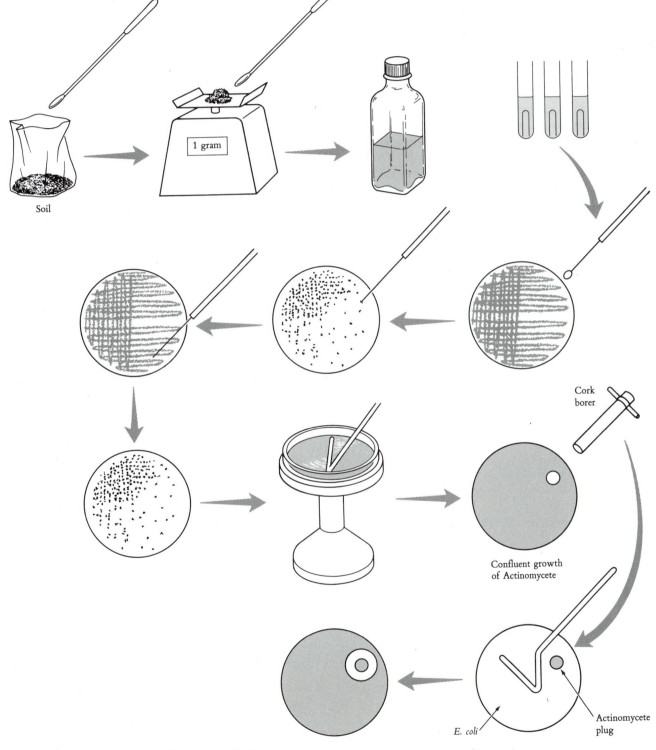

Figure 33-2 Isolating an antibiotic-secreting bacterium from soil.

at room temperature, and then refrigerate them until your next laboratory period. Identify and isolate five colonies of Actinomycetales.

Streptomyces, one of the most common antibiotic-producing Actinomycetales genus, forms small (1–10 mm) discrete lichenoid, leathery, or butyrous colonies. The colonies are initially smooth but later develop a granular, powdery, or velvety appearance as an aerial mycelium grows. Color of mature sporulated mycelium varies from white, gray, yellow, red, green, to blue-gray. Restreak to ensure purity.

4. Inoculate a plate with each actinomycete so that the entire surface will be covered by the growth.
5. For a test system, prepare two emulsion agar plates, one of *Staphylococcus aureus* and one of *Escherichia coli*.
6. After 48 hours of growth at room temperature, punch out plugs of the pure culture of each Actinomycetales species with a 1-cm alcohol-flamed cork borer and invert one plug on the surface of the hardened emulsion plates of the test organisms.
7. Incubate at 37°C for 24 hours. Record your results on the report sheet.

QUESTIONS

1. Why are there so many antibiotic-producing organisms in soils?
2. What are the characteristics of Actinomycetales? How do you recognize an actinomycete colony? How do you recognize Actinomycetales on a microscope slide?
3. Can you name one well-known antibiotic produced by an actinomycete and the organism that produces it?

ANTIBIOTIC SUSCEPTIBILITY

Illustrate the relative effectiveness of agar discs from the *Penicillium* plate in inhibiting *Staphylococcus aureus* and *Escherichia coli* by drawing zones of inhibition of appropriate sizes in the circles.

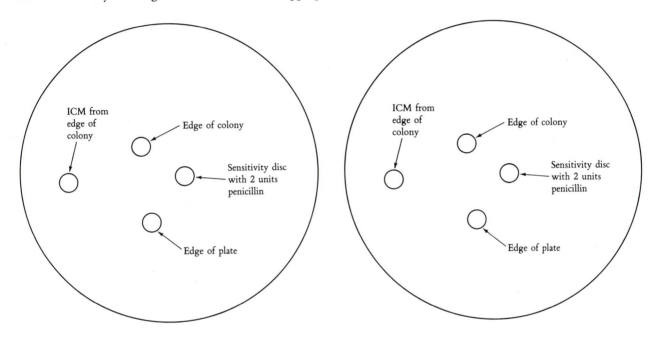

ISOLATING AN ANTIBIOTIC PRODUCER

Illustrate the effectiveness of agar discs from the Actinomycetales plate in inhibiting *Staphylococcus aureus* and *Escherichia coli* by drawing zones of inhibition of appropriate sizes in the circles.

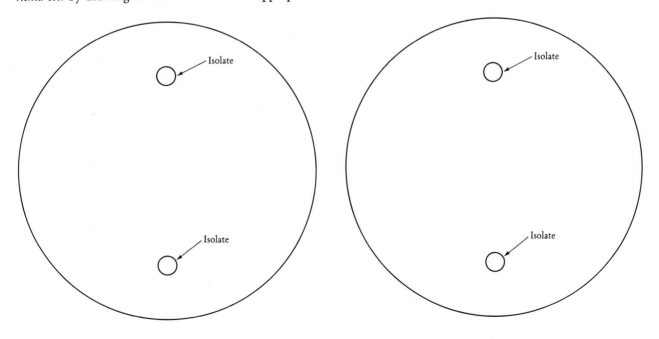

34. *Enrichment Cultures and the Winogradsky Column*

One of the most powerful tools used in general microbiology is the *selective* or *enrichment* culture. This technique was first used successfully by the Dutch microbiologist Martinus Beijerinck in the early 1900s. Beijerinck considered the enrichment culture to be a way to apply Darwinian selection to bacteria. Enrichment culture entails devising a culture medium and physical–chemical conditions that favor growth of a class of organisms of interest, or even a single species. In contrast, a nonselective medium such as nutrient agar will culture nearly all the organisms in an environment.

For example, using glucose as an electron donor under aerobic conditions enables a variety of microorganisms to grow, but a more restricted group of bacteria grows if you add methane or dichlorophenol as the sole electron donor. Similarly, elemental sulfur, used as the electron donor under aerobic conditions (with oxygen as the electron acceptor), usually selects for *Thiobacillus* and other sulfur-oxidizing bacteria. You can devise a medium with a pollutant such as Agent Orange (2,4,5-T) as sole electron donor if you are interested in enriching organisms capable of its breakdown.

Varying the electron acceptor can also select for various classes of microorganisms. Excluding air and adding nitrate will select for denitrifiers; adding sulfate selects for sulfate reducers, and using CO_2 as the sole electron acceptor selects for methanogens. Another energy source that can affect the outcome of an enrichment is light, the presence of which obviously favors phototrophs.

Lack of any carbon source other than CO_2 selects for autotrophs that fix their own carbon, and media that lack fixed nitrogen select for organisms capable of nitrogen fixation. Inoculating soil into a medium that contains glucose and no fixed nitrogen and incubating aerobically usually enriches the aerobic *Azotobacter*, whereas incubating the same medium anaerobically often selects for *Clostridium pasteurianum*. It is wise to consult the literature for possible nutrient requirements of a group of organisms you are interested in. For example, certain purple sulfur bacteria require vitamin B_{12}.

Physical and chemical factors can determine which organism predominates and can alter the effects of enrichment. Low pH in your medium favors acidophiles, and high temperature favors thermophiles. Heating soils briefly to 80°C selects for spore-formers, which can survive that treatment. Selective agents can also affect the outcome of enrichment. Adding procaryotic inhibitors such as penicillin and streptomycin favors eucaryotes. Adding cycloheximide, an inhibitor of eucaryotes, favors bacteria.

The source of inoculum is also important. It has been said that every species of microbe can be found in a single gram of soil. Indeed, many organisms that might not be expected in aerobic soils, such as some obligate anaerobes and thermophiles, are found in sufficient numbers for a positive enrichment (you need only a single viable organism). In practice, it is a good idea to use an inoculum from a habitat likely to contain the organism of interest. Thus, rumen fluid, anaerobic sewage sludge, and black pond mud are good sources of anaerobes; a toxic waste dump is a good source of bacteria with novel biodegradative abilities; and hot springs are sources of extreme thermophiles.

Perhaps by now you have inferred (correctly) that enrichment cultures are part art and part science. You can design them rationally, but subtle and unknown details, rather than theory, decide which organism turns up.

One of the classic enrichment culture techniques, which selects for several types of microorganisms that oxidize and reduce sulfur compounds, is the Winogradsky column. The basic concept is to establish opposing gradients of sulfide and oxygen in mud that is incubated in the presence of light, thereby giving an opportunity for a variety of phototrophic and chemotrophic sulfur oxidizers to establish themselves at various points along these gradients. These gradients mimic natural circumstances in which aerobic and anaerobic zones meet, such as where anaerobic muds meet aerobic pond water.

Sulfide is produced by adding cellulose to the mud as an electron donor and sulfate as the acceptor. Cellulose is not used directly by known sulfate-reducing bacteria but is hydrolized to sugars and oligosaccharides, which are then fermented to organic acids and H_2 by various fermentative bacteria such as *Clostridium*. Cellulose is fermented more slowly than a soluble sugar, which helps prevent acidification of the mud. The sulfate reducers can then use the organic acids and H_2 as electron donors to reduce sulfate to sulfide. The mud turns black due to the formation of an iron sulfide precipitate. The sulfide tends to diffuse upward, creating a concentration gradient that increases toward the bottom of the tube. Opposing this sulfide concentration gradient is an oxygen gradient that increases toward the upper part of the column where oxygen is produced by cyanobacteria and algae if present.

The green and purple photosynthetic sulfur bacteria can use sulfide as a reductant for the fixation of CO_2. Green sulfur bacteria (*Chlorobiaceae*) are generally more sulfide-tolerant than purple sulfur bacteria (*Chromatiaceae*) and are usually found below them in the column. The non-sulfur purple bacteria (*Rhodospirillaceae*) are the least tolerant of sulfide, and if sufficient organic matter is present to support their growth, they can be found in an anaerobic zone above the purple sulfur bacteria.

Where the sulfide and oxygen meet, which may be at the mud–water interface or below it, one can often find aerobic sulfur oxidizers, the so-called colorless sulfur bacteria, which oxidize the sulfide that is diffusing toward them from below by using the oxygen diffusing from above to produce sulfate. These can include the rod-shaped *Thiobacillus* and the filamentous *Beggiatoa*.

Finally, cyanobacteria and algae can develop in the waters above the mud, along with various aerobic bacteria that grow on the organic matter diffusing up from the mud, from the algae, and perhaps from protozoa that graze on the bacteria and the algae.

PROCEDURE

1. Collect a sample of water and mud from a salt marsh, pond, or lake. Remove stones and large particles from the material. In a beaker, mix the mud by volume 6:1 with $CaSO_4 \cdot 2H_2O$ and 8:1 with cellulose powder. Thoroughly mix with the aid of a spatula. Fill a 25×150 mm test tube with the mixture to about 4 cm from the top. As you add portions of the enriched mud mixture to the tube, you may want to pipet small amounts of water into the tube to displace air trapped in the mud. Pack the mud mixture down with a thick glass rod as you fill the tube. The column should be packed well

enough that no air bubbles are visible. An ungraduated cylinder may be used instead of a test tube.

2. Add some of the water sample to the mud column so that, after the column has settled, a liquid layer of about 2–4 cm covers the mud mixture.

3. Cover the top of the tube with plastic wrap or a rubber stopper to retard evaporation. The water on top of the mud mixture must not be allowed to evaporate during the incubation. Add distilled water when necessary. You may need to pack the mud down occasionally if bubble formation occurs.

4. Incubate the column at room temperature in light, either 2 feet from a 60 watt bulb or near a window that does not get direct sunlight.

Observations

Check your column *every week* and record all changes in your lab notebook. Note when the color of your column becomes darker and when gas pockets appear. Look for the appearance of green, brown, or purple patches and their locations in the column (Figure 34-1). Several weeks after inoculation, you will sample the various areas of the column by withdrawing a sample from the side of the column using a Pasteur pipet or syringe.

After removing a sample, place a drop on a slide and observe microscopically in a wet preparation. Some of the microorganisms you may find in your column are the following:

Aerobic, sulfide-oxidizing bacteria, for example, *Thiothrix* and *Beggiatoa*.

Sulfate-reducing microorganisms such as *Desulfovibrio*.

Sulfur-oxidizing bacteria that yield H_2SO_4 and thiosulfate, such as *Thiobacillus*.

The red sulfide-oxidizing, anaerobic bacteria that grow in illuminated areas, for example, *Chromatium*. Especially interesting are the purple sulfur bacteria. They contain elemental sulfur granules, which are usually multicolored under phase-contrast and light brown under bright-field. The purple bacteria are often phototactic. This can be tested by closing down the iris diaphragm of your microscope when viewing a thick suspension. If the cells are photoactic, they will accumulate in the cone of light. Note the behavior of

Figure 34-1　Typical Winogradsky column.

individual bacteria when they leave the illuminated area.

Anaerobic, sulfide-oxidizing bacteria such as *Chlorobium*, which are all photosynthetic.

Look for an area with a whitish veil where oxygen and sulfur meet in the column. Look for small rod-shaped bacteria (possibly *Thiobacillus*) or large filamentous bacteria with sulfur granules (*Beggiatoa*). Also, examine the upper water layer for cyanobacteria, algae, bacteria, and protozoa).

Make simple drawings of the various forms you see.

After your microscopic and macroscopic examination, you may want to isolate representative species from one or more of the photosynthetic groups. If so, you might use the basal medium outlined in Table 34-1, supplemented by the organic and inorganic nutrients required for the group you are isolating.

Table 34-1　Enrichment Media and Conditions for the Isolation of Photosynthetic Bacteria[a]

	Additions to Medium		Special Environmental Features		
Organisms To Be Enriched	Organic		Inorganic	Atmosphere	pH
Blue green bacteria	None		None	Air, or air + 5% CO_2	6.0–8.0
Green sulfur bacteria	None		NH_4Cl, 1.0; $Na_2S \cdot 9H_2O$, 2.0; $NaHCO_3$, 5.0	None (stoppered bottle)	7.5
Purple sulfur bacteria	None		NH_4Cl, 1.0; $Na_2S \cdot 9H_2O$, 1.0; $NaHCO_3$, 5.0	None (stoppered bottle)	8.0–8.5
Purple nonsulfur bacteria	Sodium malate, 5.0; yeast extract, 0.5		NH_4Cl, 1.0	None (stoppered bottle)	7.0–7.5

[a]The components of the medium are in grams per liter. Basal medium: $MgSO_4 \cdot 7H_2O$, 0.2; K_2HPO_4, 1.0; $FeSO_4 \cdot 7H_2O$, 0.01; $CaCl_2$, 0.02; $MnCl_2 \cdot 4H_2O$, 0.002; $NaMoO_4 \cdot 2H_2O$, 0.001; NaCl, 0.5. Environment: constant illumination, temperature 25°–30°C.

QUESTIONS

1. How do each of the three chemicals added in step 3 function in the Winogradsky column?
2. If the column were not wrapped in foil, what would be the nature of the microbial population?
3. Why does the concentration of hydrogen sulfide decrease near the surface of the column?
4. What genera of microorganisms are likely to predominate at the surface?

Enrichment Cultures and the Winogradsky Column

Record your observations of the column over a period of several weeks. Indicate color changes and visible gas formation. Indicate the morphological types of cells that you observe microscopically on successive samplings. Based on your observations, suggest groups of microbes or genera that have developed in your column. Consult *Bergey's Manual of Systematic Bacteriology* for additional information.

Enzymatic Reactions

Microorganisms, like all living things, can modify their environment to some extent and utilize chemicals in solution as sources of energy and as building blocks for growth and reproduction. All cellular activities are mediated by enzymes, and in the complex chemical reactions of life, the microorganism produces many types of enzymes whose activities interlock. By testing the chemical end products of enzymatic actions and by noting the disappearance of certain substances from the medium, microbiologists can establish the enzymatic makeup of a microorganism and can identify it and differentiate it from closely related species.

Most microorganisms, like humans, use various carbohydrates as their main source of energy. These include **polysaccharides** (complex carbohydrates), **disaccharides**, and **monosaccharides**. A variety of carbohydrates and a few other substances commonly used by microorganisms are listed in Table VIII-1.

Some microorganisms ferment many of these substances; others ferment only a few. They also vary in the way in which they break down a particular carbohydrate. Depending on the species, various end products are formed, such as organic acids (lactic, acetic,

Table VIII-1 Types of Carbohydrates

Monosaccharides	Disaccharides $(C_{12}H_{22}O_{11})$	Trisaccharides $(C_{18}H_{32}O_{16})$	Polysaccharides
Hexoses	Maltose	Raffinose	Hexosans
$(C_6H_{12}O_6)$	Sucrose	Melizitose	$(C_6H_{10}O_5)_n$
Glucose (dextrose)	Lactose		Starch
Fructose (levulose)	Cellobiose		Inulin
Galactose	Melibiose		Dextrin
Mannose			Glycogen
Sorbose			Galactan
			Cellulose
Pentoses			
$(C_5H_{10}O_5)$			Pentosans
Arabinose			$(C_5H_8O_4)_n$
Xylose			Araban
Rhamnose $(C_6H_{12}O_5)$			Xylan
Polyhydric alcohols			Glucosides
Mannitol			Salicin
Glycerol			Amygdalin
Adonitol			Esculin
Dulcitol			
Sorbitol			

butyric, and propionic), neutral products (acetone, butyl alcohol, and ethyl alcohol), and various gases (methane, hydrogen, and carbon dioxide).

The exercises that follow are intended to illustrate the diversity of microbial action on carbohydrates and some of the methods we use to study these actions.

35. *Degradation of Carbohydrates*

CARBOHYDRATE FERMENTATION

Among the common products of carbohydrate breakdown by microorganisms are organic acids (for example, acetic and lactic acids) and gases (such as carbon dioxide and hydrogen). The types and proportions of products formed depend on the species of microorganism and the carbohydrate being dissimilated. Thus, the ability (or lack of ability) of an organism to break down a single sugar, combination of sugars, or various other carbohydrates in media provides information for the classification of a bacterial species. The easy detection of acid and gas produced by individual species of bacteria is of practical importance.

You can easily detect the formation of acids by including a pH indicator in the growth medium. A commonly used pH indicator is phenol red, which is red at pH 8.5 and yellow at pH 6.9. Gas formation in broth can be detected by using a vaspar seal or an inverted vial (a Durham tube). Gas production in agar media forms gas pockets that appear as cracks in the agar.

Some microorganisms do not ferment carbohydrates but dissimilate them by oxidation. Acid formation from carbohydrate breakdown by these microbial types occurs only in aerobic conditions. In agar media in tubes, acid formation by these oxidative species occurs only at the agar surface, or not at all if the agar is layered with mineral oil or vaspar. Acid production in a covered agar culture tube indicates carbohydrate fermentation. Whether a carbohydrate is oxidized or fermented is important in identifying certain gram-negative rods, for example, the pseudomonads (see Exercise 44).

This exercise illustrates some simple methods you can use to detect acid and gas formation result-

ing from carbohydrate breakdown. In Procedure A, broth media containing the pH indicator phenol red are used to detect acid formation, and an inverted vial or a vaspar seal is used to determine gas formation (Figure 35-1).

Procedure B employs paper discs, each impregnated with a particular carbohydrate. These discs are placed on the surface of a phenol red agar plate that was previously inoculated with the test organism. The sugar in each disc diffuses into the surrounding agar, and the fermentation of a particular sugar is indicated by a yellow zone around its respective disc (Figure 35-2).

Procedure C illustrates a procedure for determinating whether a carbohydrate is oxidized or fermented. This method uses OF (oxidation–fermentation) agar containing the specified sugar and bromthymol blue as a pH indicator. Bromthy-

Figure 35-1   Reactions in phenol-red fermentation tubes.

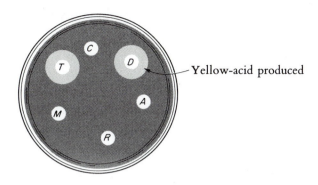

Figure 35-2 The disc method for determining carbohydrate fermentation. Letters indicate type of carbohydrate.

mol blue is purple at pH 6.8 and yellow at pH 5.2. Two tubes of OF agar are inoculated by stabbing into the agar (almost to the bottom of the tube) with the inoculating needle. The inoculated agar of one tube (the fermentation tube) is covered with a thin layer of vaspar. Fermentative cultures will turn both tubes of OF agar yellow, but oxidative cultures will turn only the top few millimeters of the uncovered agar yellow. In some oxidative cultures, no acid end products will be detected (Figure 35-3).

PROCEDURE A

1. You are provided with four tubes each of the following media:

Figure 35-3 Oxidation–fermentation tubes.

(a) Phenol-red glucose broth containing a Durham tube.
(b) Phenol-red sucrose broth containing a Durham tube.
(c) Phenol-red lactose broth containing a Durham tube.

Inoculate three tubes of each medium with *Escherichia coli, Enterococcus faecalis,* and *Proteus vulgaris.* The fourth tube of each set is a control tube and is not inoculated. *Note*: Be sure to label each tube and be sure to place the cap or plug back on the same tube it came from!

2. Using the culture of *Escherichia coli,* inoculate a tube of phenol-red glucose broth. Cover the inoculated tube with a vaspar seal.
3. Incubate all tubes at 37°C.
4. At the next laboratory period observe the cultures for acid and gas production, and record the results in the table on the report sheet.

Ecoli, B subtilis St faecalis

PROCEDURE B

1. Inoculate phenol-red agar plates with *Escherichia coli, Enterobacter aerogenes, Proteus vulgaris,* and *Pseudomonas aeruginosa.* Spread the inoculum over the entire surface of the agar plate by using sterile swabs or the bent glass rods used in spread plating.
2. Using flame-sterilized forceps, evenly distribute over the surface of each plate the six different carbohydrate discs supplied.
3. Incubate 18–24 hours at 37°C in the upright position. For evidence of carbohydrate fermentation, observe the carbohydrate discs on each plate for changes in the phenol-red indicator to yellow. Record the results in the table on the report sheet.
4. Reincubate and observe again after 48 hours. This second observation is necessary because an organism does not necessarily ferment all carbohydrates at the same rate. Furthermore, a positive sugar fermentation can be reversed by oxidation of the acid end products. Thus, an organism may show fermentation of a carbohydrate by a color zone reaction at 24 hours, and the reaction may reverse (alkaline reversion) by 48 hours when the fermentation of a second carbohydrate is just beginning.

PROCEDURE C

1. Inoculate two tubes of glucose OF agar, from the slant cultures of *Escherichia coli* and *Pseudomonas aeruginosa* provided, by stabbing once with an inoculating needle.
2. Cover one tube of each pair with a thin layer of vaspar.
3. Incubate 48 hours at 37°C.
4. Observe for production of acid and gas from glucose. Record the results on your report sheet. Fermentative cultures turn both tubes yellow, and oxidative cultures turn only the surface of the open tube yellow.

The stabbing technique also enables you to observe for motility. Motile cultures cause turbidity throughout the agar because of the movement of microbes through the soft medium, whereas nonmotile cultures grow only along the stab line.

QUESTIONS

1. Name some acids produced by bacteria. Name the principal genera that produce each acid.
2. Is the type of acid produced of value in microbial taxonomy? Explain.
3. What is meant by the terms *homofermentative* and *heterofermentative*?
4. How is the inverted Durham fermentation tube filled with medium?
5. Some organisms use carbohydrates without forming detectable levels of acid. In such a situation how would you determine if the carbohydrate was used?
6. What reaction occurs during alkaline reversion? During the oxidative dissimilation of glucose?
7. What particular characteristic of OF agar makes it useful in determining motility?

CITRATE UTILIZATION

The ability to utilize citrate as a sole source of carbon and energy distinguishes certain gram-negative rods. Simmon's citrate agar contains citrate as its only carbon and energy source. Growth on this medium is a positive test for citrate utilization. Certain organisms that give a positive test increases

green

the pH of the agar, changing the bromthymol blue indicator in the medium from green to blue.

PROCEDURE *green agar plates*

1. Inoculate a slant of Simmon's citrate agar with cultures of *E. coli, E. aerogenes,* and *P. aeruginosa.*
2. Incubate at 37°C for 48 hours.
3. Growth during incubation is a positive test.
4. Record results on your report sheet.

QUESTION

Can you name any otherwise very similar organisms that can be distinguished by their ability or inability to utilize citrate as a sole carbon and energy source?

STARCH HYDROLYSIS

Starch is a complex carbohydrate of the polysaccharide type. A qualitative test for the presence of starch is the appearance of a blue color upon addition of a solution of iodine. After starch is **hydrolyzed** (decomposed with the addition of water), the cleavage products—dextrins, maltose, and glucose, to list them in descending order of complexity—do not give this color reaction. This principle is used in testing for starch hydrolysis in this exercise.

An organism can hydrolyze starch only if it produces the enzyme **amylase**. Because not all microorganisms produce amylase, starch hydrolysis can be used to aid in identifying unknown microbes.

PROCEDURE

1. Inoculate two starch-agar plates, one with a culture of *Escherichia coli* and one with *Bacillus subtilis,* by touching the inoculating loop to the center of each plate. Because you are not attempting to isolate colonies and because you want an area of contrast around the colony after it has been developed with iodine, confine the inoculum to the centers of the plates.
2. Incubate at 37°C until the next laboratory period.

3. Test for the **diastatic** (starch-hydrolyzing) action of the organisms by flooding the surface of the starch-agar plates with gram's iodine. A clear zone around the area in which the organisms have grown indicates hydrolysis of the starch.

QUESTIONS

1. Why are amylases classed as hydrolytic enzymes?
2. Are amylases exo- or endoenzymes?
3. What is the ecological advantage of these enzymes?

METHYL-RED/VOGES – PROSKAUER TEST (MR-VP)

These tests are tools for identifying gram-negative, nonsporeforming rods and some species of *Bacillus*. Instead of accumulating mostly acidic products from fermenting glucose, some bacteria convert the metabolic intermediate, pyruvic acid, to neutral products and CO_2. Acetylmethylcarbinol ($CH_3COCHOHCH_3$) is one such neutral product that is easily detected by the Voges–Proskauer Test (Figure 35-4).

The methyl-red test detects organisms that do not convert acidic products to neutral products and

therefore remain acidic. The methyl-red indicator changes to a red color, which is a positive test.

PROCEDURE

1. Inoculate tubes containing 5 ml of MR-VP medium with *E. coli* and *E. aerogenes*.
2. Incubate for 48 hours at 37°C.
3. Test for acetylmethylcarbinol:
   (a) Decant about one-quarter of the culture into a clean test tube.
   (b) Add 0.5 ml (8–10 drops) of the alpha-naphthol solution (5% solution in alcohol).
   (c) Add 0.5 ml of the 40% KOH solution containing 0.3% creatine.
   (d) Shake thoroughly and allow to stand for 5–30 minutes.
   (e) A pink to red color indicates the presence of acetylmethylcarbinol.
4. Test for acid by adding a few drops of an alcoholic solution of methyl red to the rest of the culture. A distinct red color is a positive test; yellow is negative.

Note that the tubes contain only 5 ml of medium. This establishes a ratio of a large volume of air in the test tube to a small volume of medium, conditions that promote acetylmethylcarbinol formation.

Figure 35-4 Relation between acid and acetylmethylcarbinol production by *Enterobacter aerogenes* and *Escherichia coli* as a function of time.

Degradation of Carbohydrates

Name _____

Desk No. _____

CARBOHYDRATE FERMENTATION

PART A

Record A for acid production; G for gas production.

Inoculum	Glucose	Sucrose	Lactose
Escherichia coli			
Streptococcus faecalis			
Proteus vulgaris			
Control			

Record your observations on the appearance of the glucose broths with vaspar seal by completing the sketch.

Glucose broth

PART B

Label the carbohydrate discs and indicate which were fermented by each of the four cultures. Were the acid products of any culture oxidized during continued incubation?

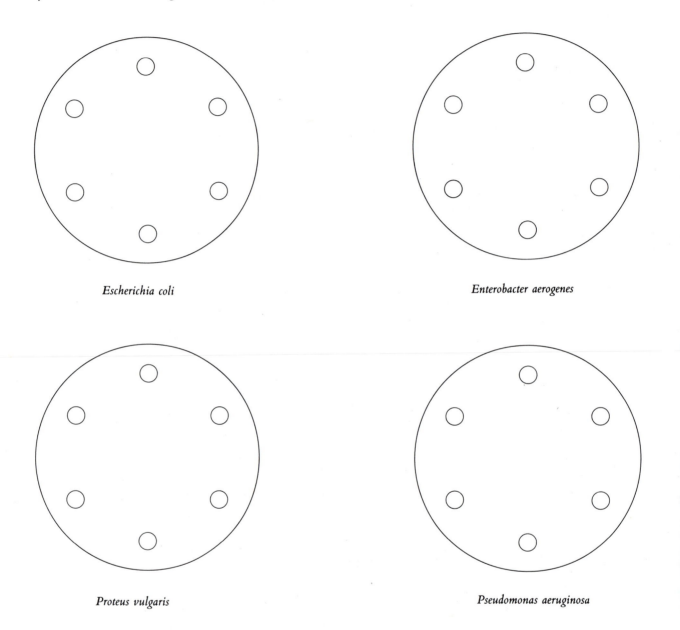

Escherichia coli

Enterobacter aerogenes

Proteus vulgaris

Pseudomonas aeruginosa

PART C

Record the reactions and describe the appearance of *Escherichia coli* and *Pseudomonas aeruginosa* on glucose OF agar.

CITRATE UTILIZATION

Record the results of the growth of the three cultures on the Simmons citrate agar.

STARCH HYDROLYSIS

Indicate areas of growth and approximate zones of starch hydrolysis.

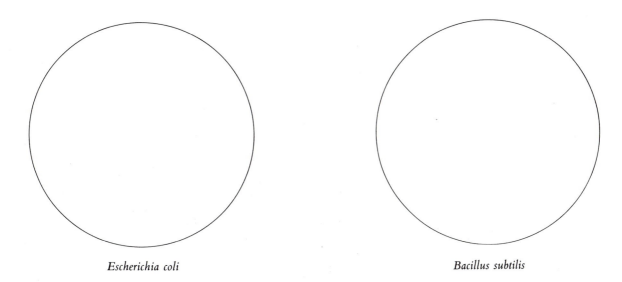

Escherichia coli   *Bacillus subtilis*

METHYL-RED/VOGES – PROSKAUER TEST (MR-VP)

Record the results of these tests and the citrate test in the following table.

Organism	Methyl Red	Voges – Proskauer	Citrate
Escherichia coli			
Enterobacter aerogenes			

Many bacteria can degrade a variety of proteins and use the resulting peptides and amino acids for energy and to synthesize their own proteins. **Proteolytic** ability varies from species to species as do patterns of amino acid dissimilation. These features can be used to characterize a species.

36. *Degradation of Proteins and Amino Acids*

CASEIN HYDROLYSIS

Casein, which exists in milk as a colloidal suspension that gives milk its opaque whiteness, is the principal protein in milk. Many bacteria produce enzymes that hydrolyze this protein into more soluble and transparent derivatives. Protein breakdown, sometimes called **peptonization**, is also useful in identifying microbial species.

PROCEDURE

1. Obtain two petri plates of skim milk agar.
2. Inoculate one plate with *Escherichia coli* and the other with *Bacillus subtilis,* using a loop to place cells in the center of the plates only (as in Exercise 35C).
3. Incubate at 37°C until the next laboratory period.
4. Illustrate your findings on the report sheet.

Colonies of proteolytic microbes (organisms that digest casein) will be surrounded by clear zones. Areas in which the casein has not been attacked will remain slightly opaque. You can see the clear zones best against a black background.

QUESTIONS

1. What derivatives are formed in casein breakdown?
2. Are the enzymes responsible for casein hydrolysis exo- or endoenzymes?

GELATIN HYDROLYSIS

The protein **gelatin** is obtained by hydrolyzing collagen, a component of connective tissue and tendons of animals. Gelatin is a good substrate for testing microorganisms for proteolytic enzymes.

Water solutions of gelatin at the concentrations used in this exercise are liquid at room temperature but solidify in an ice bath. If the gelatin has been hydrolyzed by the microorganisms being tested, the medium will remain liquid in an ice bath.

PROCEDURE

1. Inoculate four tubes of nutrient gelatin with *Escherichia coli, Bacillus subtilis, Enterococcus faecalis,* and *Proteus vulgaris.*
2. Incubate at 37°C, together with a sterile tube of nutrient gelatin as a control. Test at 2 days and up to 7 days, or until a positive reaction is obtained.
3. To examine for hydrolysis, chill the tubes in ice water. The control tube and tubes in which no hydrolysis has taken place will solidify. Hydrolyzed gelatin will remain fluid (Figure 36-1).
4. Record your findings on the report sheet.

Take care not to agitate the tubes while their contents are liquid. If hydrolysis of the gelatin is proceeding slowly from the surface downward, as it often does, agitation can mix the hydrolyzed with the unhydrolyzed gelatin and give an erroneous result.

Protein breakdown is slow compared with the breakdown of carbohydrates, and visual detection may take careful scrutiny and longer incubation.

QUESTIONS

1. Can you suggest how a test for gelatin hydrolysis might be adapted to a petri plate?
2. How does gelatin differ from most other proteins?

Figure 36-1 Testing for gelatin hydrolysis.

3. The enzymes in both casein and gelatin hydrolysis cleave the same type of chemical linkage. What is it?

UTILIZATION OF AMINO ACIDS

Although amino acids serve primarily as the basic constitutents of the many proteins that compose living organisms, they are also utilized by cells for other purposes. Amino acids can be degraded to yield energy and a variety of end products, for example, NH_3, indole, and H_2O. Amino acids can also be chemically altered to essential cell components, including other amino acids.

The similarity among gram-negative genera that are members of the family Enterobacteriaceae poses

an identification problem for the microbiologist. This diverse family ranges from *Escherichia coli*, one of the normal intestinal flora, to *Salmonella typhi*, the cause of typhoid fever. Therefore, rapid and accurate identification of these organisms is vital. Many reactions involving the degradation of amino acids are used for differentiating the Enterobacteriaceae. This exercise illustrates some of these reactions.

DECARBOXYLATION AND AMINE PRODUCTION

The **decarboxylation** of an amino acid is the splitting off of its carboxyl group to yield an amine and CO_2. The reaction can be expressed as follows:

$$\underset{\substack{| \\ NH_2 \\ \text{an amino acid}}}{R-CH-COOH} \xrightarrow[\text{(enzyme)}]{\text{decarb-oxylase}} \underset{\text{an amine}}{R-CH_2-NH_2} + CO_2 \uparrow$$

Bacterial decarboxylation can be demonstrated by showing the disappearance of the amino acid – usually a fairly complex procedure – or the formation of the amine and the CO_2. Because the reaction results in the accumulation of an amine, which is basic, decarboxylation can also be demonstrated by measuring the rise in pH.

PROCEDURE

1. For this exercise you need two tubes of lysine decarboxylase broth, two tubes of ornithine decarboxylase broth, and two tubes of decarboxylase base broth.
2. Inoculate one tube of each type of broth with *Proteus vulgaris* and the other with *Enterobacter aerogenes*. Cover all tubes with vaspar or agar caps.
3. Incubate at 37°C for 2 days.
4. Observe for gas and pH change. A rising pH, due to the accumulation of amines, changes bromcresol purple, the pH indicator in the medium, from yellow to purple (Figure 36-2 and Table 36-1). Record reactions on the report sheet.

Amino acids are decarboxylated as an initial step in degrading them for use as essential nutrients, and energy sources. The process is also used to raise the pH of the environment to counter acidic condi-

Figure 36-2 Determination of lysine decarboxylation.

tions. Enzymes that are specific for decarboxylating particular amino acids have been described that, without exception, function only if the pH of the medium is below 7.

The foul odors that often accompany proteolysis are due to the formation of volatile amines from certain amino acids.

DEAMINATION

Deamination of amino acids is the enzymatic splitting off of the amino group to yield NH_3 and usually the corresponding keto acid:

Table 36-1 Decarboxylation and Deamination of Amino Acids

Enzyme	Proteus vulgaris	Enterobacter aerogenes
Lysine decarboxylase	−	+
Ornithine decarboxylase	−	+
Phenylalanine deaminase	+	−

In this exercise, deamination of phenylalanine will be indicated by the production of phenylpyruvic acid, its corresponding keto acid. This compound forms a colored complex with ferric ions.

PROCEDURE

1. Inoculate one slanted tube of phenylalanine agar with *Proteus vulgaris* and a second with *Enterobacter aerogenes*.
2. Incubate at 37°C until the next laboratory period.
3. After incubation, test the tubes for the formation of phenylpyruvic acid by pipetting four or five drops of a 10% $FeCl_3$ solution so that it flows over the surface of the slants. A green color indicates a positive test. Record reactions on the report sheet.

INDOLE PRODUCTION

Indole is a nitrogen-containing compound that can be formed from the degradation of the amino acid tryptophan by certain bacteria. The indole test is important because only certain bacteria form indole, and it can be readily detected chemically. Thus, the degradation of tryptophan is another differentiating reaction.

Pure tryptophan is not usually used in the test medium for indole. Instead, **tryptone** (a product of digestion of certain proteins) is used as the substrate because it contains much tryptophan. The reaction by which tryptophan yields indole is as follows:

PROCEDURE

1. Inoculate two tubes of 1% tryptone broth, one with *Escherichia coli* and the other with *Enterobacter aerogenes*.
2. Incubate at 37°C until the next laboratory period. (The test should preferably be made after 24 hours.)
3. Test each tube for the presence of indole, using the Kovacs method.
   (a) To the culture fluid (about 6 ml) add 0.3 ml of Kovacs solution.
   (b) Mix well by rotating the tube between your hands. The alcohol layer will separate from the aqueous layer upon standing, and a reddening of the alcohol layer within a few minutes indicates that indole is present.
4. Record reactions on the report sheet.

HYDROGEN SULFIDE PRODUCTION

Some bacteria act on sulfur-containing amino acids to liberate H_2S. Evidence of this reaction are the aroma of rotting eggs and the blackening of certain spoiled canned foods (caused by a reaction between H_2S and the metal of the can).

Production of H_2S by bacterial cultures can be demonstrated in the laboratory if sulfide-producing cultures are grown on media containing salts of metals such as bismuth and iron. The darkening along the line of the stab is caused by the formation of the metal sulfide.

One example of H_2S production is shown by the following reaction:

$$CH_2-SH$$
$$H_2N-CH-COOH$$
cysteine

$+ 2H$

$$CH_3$$
$$H_2S + H_2N-CH-COOH$$
hydrogen   alanine
sulfide

The sulfide–indole–motility (SIM) medium used in this exercise enables you to measure sulfide production, indole formation, and motility.

PROCEDURE

1. Using a needle, prepare stab inoculations (see Figure 8-1) of *Escherichia coli* and *Proteus vulgaris* in two tubes of SIM medium.
2. Incubate the stab cultures at 37°C for 48 hours.
3. Observe for H_2S production and motility; test for the formation of indole.

   A black precipitate of FeS indicates the presence of H_2S. The diffusion of growth away from the stab in this semisolid medium indicates the presence of motile cells. The presence of a red color in an overlay of Kovacs solution indicates the production of indole.

   The SIM medium is a good example of multiple tests in a single tube of medium.
4. Record reactions on the report sheet.

QUESTIONS

1. What is the name of the amine formed by the decarboxylation of tyrosine? Of lysine? Of ornithine?
2. What B vitamin is a component of the coenzyme in many amino acid decarboxylases?
3. What pH change would take place with deamination? Would it be as much of a pH change as occurs with decarboxylation? Why?
4. Tryptophan also reacts with *p*-dimethylaminobenzaldehyde in the Kovacs reagent. Why then is this reagent useful to detect indole production from tryptophan?
5. Could you test for H_2S production in an appropriate medium by looking for gas accumulation in a Durham fermentation tube?

ACTION OF BACTERIA ON MILK

Skim milk is an excellent medium for supporting the growth of bacteria because it contains the sugar lactose, the protein casein, and vitamins, minerals, and water. The end result of the action of bacteria on milk depends primarily on whether the organism attacks its carbohydrate or its protein. Some bacteria ferment the lactose, others proteolyze the casein, and still another group attacks both. Because a species of bacteria attacks skim milk in its characteristic ways, milk reactions are useful in identification.

Acid Production. The fermentation of the lactose in skim milk produces acid, which lowers the pH of the medium. A pH indicator such as litmus, can be added to the milk to detect acid production. As acid accumulates, the litmus color will turn from lavender to pink.

As the acidity of the milk increases, and pH falls below the isoelectric point of the casein, the casein coagulates into an acid curd (Figure 36-3A). If the acid production is rapid, the curd formation is accompanied by the squeezing of whey from the curd. Whey appears as a somewhat clear, opalescent liquid on the surface of the acid curd.

Reduction. In addition to being a pH indicator, litmus is an oxidation–reduction indicator. If oxygen is removed from the milk by the reducing action of the bacteria, the litmus is reduced to a leuco (white) state. This reduction is typified by a white zone that begins at the bottom of the tube and proceeds upward toward the surface of the acid curd. Some organisms completely reduce the litmus, and all of the litmus milk becomes white except for a rim at the surface of the curd. This surface rim is caused by the oxidation of litmus by atmospheric oxygen. In its reduced form, litmus no longer functions as an acid–base indicator.

Rennet Curd and Proteolysis. Many microbes possess a rennet-type enzyme that coagulates casein without acid production (Figure 36-3B). Many microbes that form rennet curd are also proteolytic and therefore produce enzymes that catalyze the digestion of the rennet curd. Proteolysis degrades the curd to soluble products that appear as a straw-colored fluid.

Although proteolysis usually occurs in neutral conditions, some microbes, such as *Enterococcus faecalis*, can bring about proteolysis in an acid curd.

Alkaline Reactions. Alkaline conditions arise from the decarboxylation or deamination of the amino acids of the casein, indicated by the change of litmus from lavender to blue.

Gas Formation. Some microbes produce gas during growth in skim milk, forming bubbles or pockets in the curd, usually an acid curd (Figure 36-3C). Some microbes produce so much gas that the curd is furrowed or torn to shreds. This is known as the stormy fermentation of milk and is characteristic of some clostridia and other strongly fermentative organisms.

PROCEDURE

1. Inoculate *Escherichia coli, Lactococcus lactis, Bacillus subtilis,* and *Proteus vulgaris* into tubes of litmus milk.
2. Incubate at 37°C and observe the changes that have taken place after 2 and after 7 days.
3. Record reactions on the report sheet.

QUESTIONS

1. Reduced litmus milk remains colored at the surface. Why?
2. At what pH does an acid curd form? What is the relation of this pH to the isoelectric point of casein?
3. What criterion would you use to distinguish between whey formation and proteolysis?

Figure 36-3 Action of bacteria on milk. *A:* Acid curd. *B:* Coagulation of casein without acid. *C:* Bubbles in acid curd.

Degradation of Proteins and Amino Acids

Name _____

Desk No. _____

CASEIN HYDROLYSIS

Indicate colonies and approximate areas of casein hydrolysis.

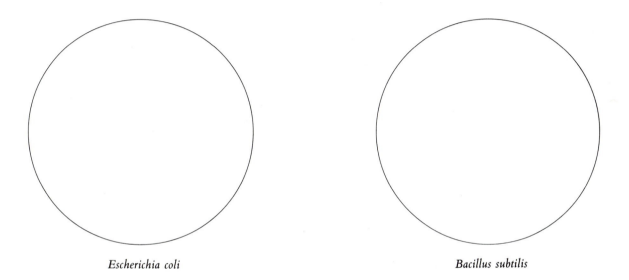

Escherichia coli   *Bacillus subtilis*

GELATIN HYDROLYSIS

Record a positive test for gelatin hydrolysis (liquid in ice bath) as + and a negative test (solid state in ice bath) as −.

Organism	Gelatin hydrolysis
Escherichia coli	
Bacillus subtilis	
Streptococcus faecalis	
Proteus vulgaris	

UTILIZATION OF AMINO ACIDS

Record the reactions of the cultures below.

Medium	Proteus vulgaris	Enterobacter aerogenes	Escherichia coli
Lysine decarboxylation			////
Ornithine decarboxylation			////
Phenylalanine decarboxylation			////
Indole production	////		
Hydrogen sulfide production		////	

ACTION OF BACTERIA ON MILK

Record your results, using AC and RC to designate acid and rennet curds, and check marks to indicate litmus reduction and proteolysis.

Culture	Acid	Coagulation	Reduction	Proteolysis
Escherichia coli				
Streptococcus lactis				
Bacillus subtilis				
Proteus vulgaris				

37. *Lipids and Urea Hydrolysis*

PHOSPHOLIPID HYDROLYSIS

The hydrolysis of phospholipids, such as those in egg yolk, is a diagnostic tool employed in microbiology to characterize or identify members of the genera *Pseudomonas, Staphylococcus, Bacillus,* and *Clostridium.* Cleavage of the phosphate ester bonds forms a water-insoluble lipid. This enzymatic activity is detected by a zone of opalescence in the medium around the cell mass (Figure 37-1).

PROCEDURE

1. You are provided with an egg yolk agar plate. Mark the bottom of the plate into quadrants with a wax pencil.

2. With an inoculating loop, inoculate a drop of the broth cultures of *Pseudomonas aerugin-*

Figure 37-1 The egg yolk reaction: demonstration of phospholipid hydrolysis.

Two lipids that are commonly decomposed by microorganisms are the **triglycerides** and the **phospholipids**. Triglycerides are esters of glycerol and fatty acids. Their hydrolysis occurs as follows:

The decomposition of triglycerides by microorganisms causes rancidity in foods that contain a high proportion of fats, for example, butter and margarine (the fatty acids formed cause the rancid flavor and aroma). One of the major problems of sewage disposal is also due to the decomposition of fats. Their presence markedly increases competition for the available oxygen and thus slows the main process of sewage decomposition. Most methods for the disposal of bulk wastes employ various methods for removing fat and grease.

The phospholipids are major components of all cell membranes. They are glycerol esters of two fatty acid residues and phosphate that are esterified to a nonfatty unit, for example, choline. Because phospholipids are functional components of all cells, the ability to hydrolyze host-cell phospholipids is an important factor in the spread of virulent microorganisms.

osa, Staphylococcus aureus, Bacillus cereus, and *Escherichia coli* into the center of each quadrant.

3. Incubate at 37°C for 48 hours.
4. Observe for an opalescent precipitate and record the results on your report sheet.

QUESTIONS

1. How is the ability to attack phospholipids associated with pathogenicity?
2. How do phospholipases A and B differ from phospholipases C and D?
3. Microbial hydrolysis of fat is primarily caused by aerobic organisms in aerobic conditions. Why?

THE UREASE TEST

The genus *Proteus* may be distinguished from some other gram-negative rods by its ability to produce large amounts of the enzyme urease. The hydrolysis of urea by urease releases NH_3, as follows:

$$\underset{NH_2}{\overset{NH_2}{C=O}} \xrightarrow[H_2O]{\text{urease}} 2\,NH_3 + CO_2 \uparrow$$

Because the production of NH_3 raises the pH of the medium, urease activity can be detected by the change of the phenol-red indicator to bright pink.

PROCEDURE

1. Inoculate a tube of urea broth with *Proteus vulgaris* and a second tube with *Escherichia coli*.
2. Incubate along with a control tube at 37°C for 24 hours.
3. The presence of a purple color is a positive test. Record the results on your report sheet.

Lipids and Urea Hydrolysis

Name _____

Desk No. _____

Indicate for each culture whether phospholipid was hydrolyzed.

Pseudomonas aeruginosa _____

Staphylococcus aureus _____

Bacillus cereus _____

Escherichia coli _____

Indicate for each culture whether urea was hydrolyzed.

Proteus vulgaris _____

Escherichia coli _____

Energy is transformed and released to meet cellular needs by oxidation, which is accomplished primarily by removing hydrogen and electrons. The hydrogen is removed from a substrate (the substance oxidized, usually a carbohydrate) and transported via various respiratory enzymes to a final hydrogen acceptor (such as oxygen). Hydrogen and electron transport supplies energy to the cell. Hydrogen is transferred and energy is conserved by a series of oxidation–reduction reactions.

Respiration is biological oxidation that occurs with atmospheric oxygen (**aerobic respiration**) or with inorganic oxides such as nitrates or sulfates (**anaerobic respiration**) as the final hydrogen acceptors. **Fermentation** is biological oxidation that occurs when the initial hydrogen donor (usually a carbohydrate) is broken down and one or more of the organic dissimilation products serve as the final hydrogen acceptor.

Whether a cell respires or ferments usually depends on the availability of atmospheric oxygen and the presence of the enzymes necessary to reduce it. **Obligate aerobes** require atmospheric oxygen as the final hydrogen acceptor and grow only in its presence. One reason for the oxygen intolerance of the **obligate anaerobes** is that they lack enzymes essential for reducing atmospheric oxygen. The relationship of microorganisms to atmospheric oxygen determines the areas of the human body that a pathogen infects. For example, obligate anaerobic pathogens infect only deep wounds and other areas where the normal blood (and oxygen) supply has been altered.

38. *Respiratory Enzymes*

CATALASE ACTIVITY

Catalase, an enzyme produced by most bacteria, catalyzes the breakdown of hydrogen peroxide to release free oxygen:

$$2H_2O_2 \xrightarrow{\text{catalase}} 2H_2O + O_2$$

In many cases gas can be seen as a white froth if a few drops of 3% H_2O_2 are added to a microbial colony or a broth culture. In the event of a questionable or apparently negative reaction, you can put a bit of culture on a slide, focus it under the low-power objective of a microscope, and add peroxide to the culture as you observe it.

Most microbial cultures taken from the usual media show unmistakable catalase reactions. Catalase-negative organisms tend to be anaerobic. Important catalase-negative genera are *Streptococcus, Leuconostoc, Lactobacillus, Clostridium,* and *Mycoplasma.*

The enzyme catalase contains the hemeporphyrin structure in Figure 38-1. This porphyrinring structure is characteristic not only of catalase but also, with Mg in place of Fe, of the **cyto-**chromes, respiratory electron carriers found in aerobic forms of life and in the chlorophyll of all photosynthetic cells.

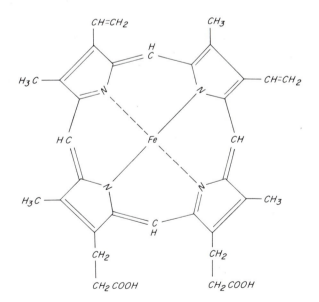

Figure 38-1 Heme, an example of an iron chelate porphyrin.

PROCEDURE A

1. Inoculate a yeast-extract slant and a tube of yeast-extract broth with *Streptococcus faecalis*. Inoculate another slant and broth tube with *Staphylococcus aureus*.
2. Incubate at 37°C for 24–48 hours.
3. Add a few drops of 3% H_2O_2 to the slant and broth cultures, and observe the broth closely for oxygen bubbles and the slant for froth on the surface. Record the results on your report sheet.

PROCEDURE B

1. Streak one blood-agar plate with *Enterococcus faecalis* and a second with *Staphylococcus aureus*.
2. Incubate at 37°C for 18 hours.
3. Dip a glass capillary tube in the beaker of 3% H_2O_2 provided, and allow capillary action to fill the tube to a height of about 20 mm. Then touch the capillary tube to a colony of *Enterococcus faecalis* on the blood-agar plate. Observe for bubbles in the capillary tube. Use a second capillary tube to repeat the procedure on a colony of *Staphylococcus aureus*. Effervescence will occur immediately in the tube touched to a colony with strong catalase activity (Figure 38-2).
4. Add a drop of 3% H_2O_2 to an area of the blood-agar plate where no colonies are present, and observe for evidence of catalase activity (Figure 38-2). Record the results on your report sheet.

The capillary tube technique is a quick and sensitive miniaturized method for detecting catalase activity. It has the added advantage of enabling you to determine the catalase reaction of growth on a medium such as blood agar that is itself catalase positive.

QUESTIONS

1. The catalase test is especially valuable in distinguishing between the gram-positive cocci _____ and _____ and
   genus   *genus*
   in distinguishing between the gram-positive long rods _____ and _____ .
   genus   *genus*
2. What is the function of superoxide dismutase?
3. Why is the capillary tube technique used to assess catalase activity of cultures grown on blood-agar plates?

OXIDASE TEST

The oxidase determines whether a microbe can oxidize certain aromatic amines, for example, *p*-aminodimethylaniline, to form colored end products. This oxidation correlates with the cytochrome oxi-

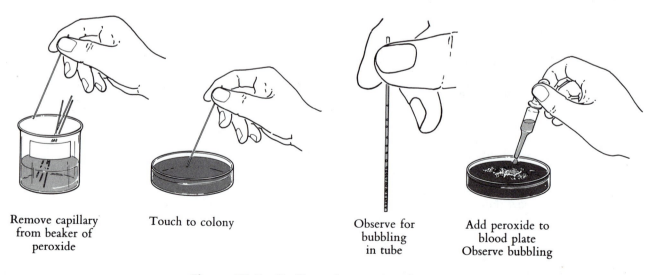

Remove capillary from beaker of peroxide   Touch to colony   Observe for bubbling in tube   Add peroxide to blood plate Observe bubbling

Figure 38-2 Capillary tube procedure for catalase test.

dase activity of some bacteria, including the genera *Pseudomonas* and *Neisseria*. While a positive oxidase test is important in the identification of these genera, the test is also useful in characterizing the enteric bacteria (Enterobacteriaceae),which are oxidase negative.

In this exercise, a test strip impregnated with the oxidase reagent is smeared with bacterial culture. In a positive test, oxidized cytochrome *c*, formed by cytochrome oxidase activity, oxidizes *p*-aminodimethylaniline to a red product.

Figure 38-3 Use of the test strip in demonstrating the oxidase reaction.

PROCEDURE

1. You are provided with a test strip and agar plate cultures of *Pseudomonas fluorescens* and *Escherichia coli*. Using a sterile loop, pick growth from a well-isolated colony on a *Pseudomonas* plate and thoroughly rub it into area 1 of the test strip (Figure 38-3).
2. After approximately 30 seconds, observe the inoculated area for a color change. A blue color indicates a positive oxidase test.
3. Using a second test strip, repeat the test with growth from a colony on the *Escherichia coli* plate. Indicate reactions on your report sheet.

Traditionally, the oxidase reaction has been determined by using an acidified solution of *p*-aminodimethylaniline. This solution is tedious to prepare and highly unstable in the aqueous state, often giving false positive reactions. However, the dry oxidase reagent on the test strip is stable. This is but one example of the advantages of using impregnated test strips in microbiological tests. Impregnated test strips can be adapted for use in many diagnostic and microbiological tests. They are better than the traditional methods because they are quick and easy to use as well as versatile and accurate.

QUESTIONS

1. What organisms process cytochrome *c* oxidase?
2. What other enzymes can give a positive oxidase test?
3. How does the reagent function in this test?
4. Why must the oxidase reaction be read immediately?

REPORT **38**

Respiratory Enzymes

CATALASE ACTIVITY

PART A

Indicate which cultures gave positive catalase reactions.

	Yeast-extract slant	Yeast-extract broth
Streptococcus faecalis	_____	_____
Staphylococcus aureus	_____	_____

PART B

Indicate which cultures gave positive catalase reactions.

Streptococcus faecalis on blood agar _____

Staphylococcus aureus on blood agar _____

Blood-agar medium _____

OXIDASE TEST

Indicate positive or negative oxidase reactions.

Pseudomonas fluorescens _____

Staphylococcus aureus _____

N I N E

Isolating and Identifying Bacterial Cultures

The study of the interrelationships of microorganisms and their environments is called **microbial ecology**. To study a particular influence on a pure culture in the laboratory, we must alter one variable at a time. But in nature, many influences are exerted on microbes simultaneously, and their interactions determine the fate of the organisms. For example, the sensitivity of an organism to heat depends to a large extent on the pH and the hydrostatic pressure of the environment. Similarly, the destructive effect of ionizing radiation depends on the presence or absence of atmospheric oxygen and other factors.

The goal of the study of microbial ecology is to understand the microorganism in relation to all features of its habitat: physical, chemical, and biological. In a practical sense, the ultimate control of microorganisms in medicine, fermentation, and the food industries will hinge on this understanding.

Habitats only rarely contain a single species of microbe. To determine what microbes live in a habitat, every species present must be isolated in a pure culture and identified. This task can often be simplified by testing for particular characteristics of the organisms: nutritional or temperature requirements, metabolic products, and so on. Thus, a thermophilic aerobe might be separated from many other microorganisms in a soil habitat by incubating a soil sample in a shallow layer of medium at 60°C. Incubating the sample in anaerobic conditions would probably grow a thermophilic anaerobe. The microbes that cannot meet the imposed conditions do not grow.

Adding selective chemical substances to the basic medium or imposing other special conditions during growth of a culture to repress the growth of interfering microbes while permitting growth of the culture sought is called **repression selection**. Imposing conditions that encourage the growth of a microbe so that it will outgrow its competitors is called **enrichment selection**. Sometimes a microbe can be isolated simply by making it easier to recognize. This procedure, which does not necessarily employ any special inhibitory or enrichment substance, is called **differential plating**.

A great many isolation procedures have been devised from various combinations of these principles. Obviously, the more information you have about the properties of a microorganism, the greater chance you have to devise a medium that will weed out competing microbes.

USING SELECTIVE ENRICHMENT CULTURES

Influenced largely by the Dutch scientist Beijerinck, and emulating nature, the bacteriologist has found it advantageous to impose particular combinations of environmental conditions on mixed flora, with the goal of favoring the development of a microorganism by giving it an advantage over its competitors. For example, adjusting the pH of a medium to about 4 prevents the growth of most bacteria and promotes the growth of acid-tolerant flora such as yeasts and molds. If a medium is prepared with sodium lactate as the sole energy source, the stage is set for the preferential development of lactate-utilizing organisms such a *Propionibacterium* or *Veillonella*. Excluding nitrogenous compounds from a medium makes the development of any but nitrogen-fixing organisms impossible. Excluding oxygen from the environment prevents the growth of aerobes. If both oxygen and nitrogenous compounds are excluded (but nitrogen gas is supplied), only the anaerobic nitrogen fixers develop.

The number of combinations of selective conditions is great, and we can isolate an organism that thrives in response to almost any set of conditions, providing, of course, that the organism is present in the sample. However, more than one type of organism usually thrives in any selective environment, and cultures so obtained are seldom, if ever, pure in the initial selective medium. Pure cultures can often be secured, however, by plating or streaking from an initial liquid-enrichment medium onto solid or semisolid medium. Sometimes, before plating, it is expedient to pass a culture through two or more transfers in a liquid-enrichment medium. This eliminates competing microbes that survive through a few generations on nutrients that were in the original sample. If, for example, a sample of rumen contents or of soil is being used as a source for an organism, the introduction of the sample to a selective medium alters the composition of the medium so that it is no longer strongly selective. Transferring the developing culture to a second flask of selective medium dilutes the contaminating nutrients. Because the organism of interest is often a minority of microbial population, the introduction of other microorganisms cannot be eliminated merely by reducing the size of the sample.

If it has not already occurred to you, consider that there is no one set of conditions that is *not* selective. Any culture is subject to many, perhaps inadvertent, selective influences: the composition of the medium, temperature of incubation, pH, oxygen relationships, and so on. As a rule, media used in refined selective or enrichment techniques more nearly approach the minimal nutritional requirements and optimum cultural conditions of a species. In this way you can limit and define the type of growth more precisely than if you include complex organic materials that invite the development of a heterogeneous population.

You can turn the selection technique in another direction by attempting to isolate an organism that is able to carry out a particular reaction. Assume, for example, that you want to isolate an organism that can hydrolyze urea. The sample, perhaps rumen contents, could be plated on a medium containing urea and a pH indicator. Of the developing colonies, those that form ammonia from urea would be shown by a change in indicator color. You could make isolations from these colonies, ascertain the purity of the cultures, and test the organism again on urea. If you want to secure a cellibiose-fermenting species, you might provide this sugar as the only energy source in a liquid medium and plate out the resulting population on cellibiose agar with an indicator to show acid production by cellibiose-fermenting colonies. In setting up any medium, you must pay careful attention to other details, for example, buffer concentration. In the preceding examples, too high a concentration of buffer would mask both ammonia and acid production.

Natural reservoirs of microorganisms such as soils, muds, and pond waters are probably the best sources of organisms not known to exist in more limiting habitats. For many species, of course, you would go to a more specific source such as sour raw milk for *Streptococcus lactis*, Swiss cheese for *Propionibacterium*, unpasteurized draught beer or fruits for *Acetobacter*, feces for *Streptococcus faecalis*, and the outer leaves of cabbages for *Leuconostoc*.

Generally, a gram or two of sample is used. Liquid media are put into small bottles or flasks, in shallow layers for aerobes or filled to the top and stoppered for anaerobes. Special atmospheres can be introduced directly into stoppered flasks or into larger containers such as desiccator jars, in which the flasks can be sealed during incubation.

If you are interested in making such a selective and enrichment culture for one of the types of bacteria noted here, discuss the possibility of such a project with your instructor.

39. *Differential Plating*

In this exercise you will take advantage of physiological characteristics of bacteria in such a way that colonies can be recognized directly on the plates. First you will be looking for lactose-fermenting bacteria in milk. By incorporating a pH indicator into the lactose agar, you can visually detect those colonies that are producing acid from lactose.

Then you will look for casein-hydrolyzing bacteria in soil, employing the same principle that you used in Exercise 36 (Casein Hydrolysis) to measure casein hydrolysis.

PROCEDURE

Lactose-Fermenting Bacteria

1. Melt three tubes of lactose yeast-extract agar containing BCP indicator. Cool then to 45°C.
2. Plate out the sample of milk that is provided, by the loop-dilution method (1, 2, and 3 loopfuls; see Exercise 9).
3. Incubate the plates at 37°C until the next laboratory period.
4. Examine the plates for colonies that are surrounded by yellow zones. The indicator BCP is yellow at pH below 5.2; thus, a yellow zone is indicative of acid formation.

Casein-Hydrolyzing Bacteria

1. Melt three tubes of skim-milk agar and cool to 45°C.
2. Plate out, by the loop-dilution method, a sample of soil suspension provided.
3. Incubate the plates at 30°C until the next laboratory period.
4. Examine the plates for colonies surrounded by clear zones. Casein, which is opaque in the plate before incubation, is converted to soluble products upon hydrolysis. Thus, a clear zone around a colony is indicative of casein digestion.

Observations

Make gram stains of three colonies showing lactose fermentation and three showing casein hydrolysis. Do the organisms from each habitat share common morphological characteristics as well as a physiological one?

QUESTIONS

1. How would you go about isolating the bacteria responsible for causing the rancidity in butter?
2. Bacteria that form acetic acid can be identified by streaking the cultures on media containing $CaCO_3$ in suspension. What is the principle of this isolation technique?

Differential Plating

Assume that the loop used in making the dilutions carried 0.01 ml of sample and that the dilutions were made in 10 ml agar deeps.

What was the number of lactose-hydrolyzing bacteria per milliliter of milk?

What was the number of casein-hydrolyzing bacteria per milliliter of soil suspension?

40. *Selective Plating*

From a sample containing a mixture of microorganisms, you can select certain species simply by choosing an appropriate repression-selection medium for plating. In the first part of this exercise you will use an antibiotic to bring about the repression selection of microorganisms from a sample of fermenting fruit juice. In the second part, you will isolate different types of bacteria from a fecal sample by plating on both azide and eosin methylene-blue (EMB) agar.

PROCEDURE

Actidione Agar

1. Melt two tubes of glucose agar (15 ml each).
2. Cool to 45°C and use one tube to pour a plate for streaking.
3. To the other tube aseptically add 1 mg (0.1 ml of the stock solution) of actidione, mix well, and pour a second plate. *Caution: Actidione is a poison; do not pipet by mouth!*
4. Streak the two plates with the fermenting fruit juice provided.
5. Incubate the plate at 30°C until the next laboratory period. Note the way the fungal growth spreads over the plate that lacks actidione and interferes with the isolation of bacterial colonies.

Antibiotics have been used as inhibitory substances in plating media with varying degrees of success. Molds and yeasts can be especially troublesome because they can overgrow the surface of a plate very rapidly and make isolations difficult. Actidione, an antibiotic produced by *Streptomyces griseus*, has proved to be valuable in suppressing the growth of many fungal species.

Note: Actidione is prepared by filtration in a sterile stock solution of 10 mg per ml of distilled water. It can be autoclaved at 122°C for 15 minutes without serious loss of potency. It can be used effectively with other selective inhibitors such as penicillin, polymyxin, and brilliant green dye; and it is not appreciably inhibitory to bacteria at the concentration used.

Azide and EMB Agar

1. Melt three tubes of tryptone yeast-extract agar containing 0.02% sodium azide and three tubes of EMB agar. Cool to 45°C.
2. Using the loop-dilution technique, plate out on both media the fecal sample provided.
3. Incubate the plates at 37°C for 2 days. Note the striking difference in the colonies that have developed on the two media.
4. With the inoculating needle pick two of the pinpoint colonies from an azide plate and two typical *E. coli* colonies (use Table 40-1) from an EMB plate into four tubes of tryptone yeast-extract broth.
5. Incubate the four tubes of tryptone yeast-extract broth at 37°C. Make gram-stain preparations and note the morphology of the cultures from the two media.

   Note: You may not always obtain pure cultures by isolating a colony from a plate of selective medium. Some microorganisms are only inhibited on the selective medium and grow when transferred to a noninhibitory medium.

Sodium azide is poisonous for iron-containing systems. Microbes that do not have these systems are relatively resistant to this substance.

EMB agar, commonly used in the detection of fecal contamination of water supplies, is both a selective and a differential medium. It is selective because certain gram-negative species (for example, *E. coli*) are more tolerant of the eosin methylene-blue complex than gram-positive species are. It is differential because of the presence of the sugar lactose. Colonies of lactose-fermenting species are pigmented, often with a metallic sheen, whereas nonfermenting colonies are colorless. This medium is of particular value for differentiating and characterizing the gram-negative nonsporeforming rods *Escherichia coli* and *Enterobacter aerogenes*. Although both of these species grow on EMB agar, as shown in Table 40-1, the appearance of colonies of the intestinal bacterium *E. coli* on this medium can be easily distinguished from those of *E. aerogenes*, a

Table 40-1 Differentiation of *Escherichia coli* and *Enterobacter aerogenes* Grown on Eosin Methylene-Blue Agar[a]

Characteristic	Escherichia coli	Enterobacter aerogenes
Size	Well-isolated colonies are 2–3 mm in diameter.	Well-isolated colonies are larger than those of *E. coli*; usually 4–6 mm in diameter or more.
Confluence	Neighboring colonies show little tendency to run together.	Neighboring colonies run together quickly.
Elevation	Colonies are slightly raised; surface is flat or slightly concave, rarely convex.	Colonies are considerably raised and markedly convex; the center may drop precipitately.
Appearance by transmitted light	Dark, almost black centers that extend across more than three-fourths of the diameter of the colony; internal structure of central dark portion is difficult to discern.	Centers are deep brown, not as dark as those of *E. coli* and smaller in proportion to the rest of the colony. Striated internal structure is often observed in young colonies.
Appearance by reflected light	Colonies are dark, buttonlike, often concentrically ringed with a greenish metallic sheen.	Much lighter than *E. coli*; metallic sheen is not observed except occasionally in a depressed center.

[a]Modified from Levine, 1921. Iowa Eng. Exp. Sta. Bull. No. 62.

closely related species whose usual habitats are soil and plants.

QUESTIONS

1. From what microbial habitats other than fermenting fruit juice might you have obtained material that would give results similar to those obtained in the first part of the experiment? In place of feces in the second part?

2. Media containing azide or cyanide should not be used in making the catalase test. Explain.

Selective Plating

Name _____

Desk No. _____

Make drawings of typical organisms from EMB agar.

Make drawings of typical organisms from sodium azide agar.

IDENTIFYING DIFFERENT GROUPS OF BACTERIA

Taxonomy is the science that attempts to describe and classify organisms into categories that reflect degrees of relatedness and degrees of distinctiveness.

Microorganisms do not have a variety of definite anatomical features and obvious patterns of evolutionary relationships such as those that the botanist and zoologist use in taxonomic work. The identification schemes for bacteria have, therefore, been based on a variety of characteristics that include not only morphology, but also cultural characteristics, physiology, pathogenicity, serology, and, more recently, molecular information. As the biologist uses anatomy in the classification of higher plants and animals, microbiologists use cell shape, cell arrangement, and cell structure. The appearance of cultures on laboratory media – colony shapes and pigmentation – is also a valuable taxonomical tool. The abilities of an organism to utilize or attack certain substances (substrates), to produce changes and chemical products that can be tested for or recognized, and to cause a particular disease are all important for the differentiation bacteria. A supplementary but powerful tool in identifying species of groups such as the *Streptococcus* and *Salmonella* is the serological test, the theory and technique of which you will learn later.

On a large scale, the nonphotosynthetic bacteria can be separated into autotrophs and heterotrophs. The autotrophs use carbon dioxide and inorganic salts as sources of carbon and nitrogen, oxidize inorganic compounds for energy, and are quite independent of organic matter for survival. They are unimportant as agents of spoilage or disease but do play exceedingly important roles in some elemental cycles in nature. In general, the nonphotosynthetic autotrophs grow more slowly than heterotrophs and are less easily identified in the laboratory.

The heterotrophs secure carbon and energy from organic materials, and many depend on organic sources of nitrogen. This group, which includes most bacteria of medical and commercial importance, live in many ways in many habitats.

In spite of the great variability of living cells and their proneness to mutation, it must be assumed and it is important to recognize the fact that *in natural conditions an equilibrium exists between the cells of a species and their environment that favors the continuation of a given species in that environment.*

In essence, that assumption is the basis of our classification scheme: the assurance, in spite of species variability, that certain microorganisms will be found in certain habitats and that their reactions are consistent enough to permit identification. It is true that an old laboratory strain may lose some of its distinguishing features. *Streptococci*, for example, may lose their ability to ferment certain sugars after being carried for a long time on artificial laboratory media. Likewise, a typhoid organism may become avirulent, or a *Pseudomonas* may no longer form green pigment. Changes occur that make an organism atypical by a rigid description. And indeed many atypical organisms often can be isolated from nature, but the student who is experienced in taxonomy recognizes that variation among individuals and populations is the rule among living species. Thus, although it is convenient to use certain easily recognized traits for diagnosis, one must acknowledge that single characteristics are usually not a solid foundation for taxonomic separation. This fact has been recognized by increasing numbers of microbiologists in recent years and underlie the trend toward numerical approaches to taxonomy, approaches which compare similarity of many different characteristics.

A test that is important for distinguishing one species or genus may be worthless for another. The fact that *Escherichia coli* ferments lactose with acid and gas production is important in distinguishing it from other short gram-negative rods. Yet among other groups of bacteria, this particular reaction is of little or no use in identification. Likewise, among the cocci, *Streptococcus* can be distinguished from *Staphylococcus* because the latter produces catalase and the former does not. In distinguishing among species of the *Escherichia – Enterobacter* group, however, the test is worthless because the organisms all produce catalase.

At this time it is appropriate to mention *Bergey's Manual of Systematic Bacteriology.* These volumes (of encyclopedic proportions) embrace a system of taxonomy that is, by and large, followed by most American bacteriologists. In addition to placing the individual species in the larger classification of genus, family, and order, where possible, there are detailed cultural descriptions of many hundreds of species of bacteria.

The task of the taxonomist is made most difficult because many fine descriptions from past years are virtually impossible to fit into the present-day scheme because culture techniques have changed. Many of the criteria that are now used to define a species or a genus were not used earlier, and some criteria used previously have now been abandoned. Yet, in justification to these pioneer workers, this material cannot be entirely disregarded and must be used when possible in establishing priority in naming or describing a species.

In many cases, the information available to the taxonomist about neglected groups of bacteria is insufficient to set up a working scheme, and classification of some groups must await further study.

41. *Identifying Unknown Bacterial Cultures*

One of the most fundamental problems in bacteriology is recovering, purifying, and identifying pure cultures of bacteria. All identification procedures are based on the assumption that a test culture is pure; therefore, we cannot overemphasize the importance of making sure a culture is pure to begin with and of keeping it free of contamination.

In this exercise you will be given cultures to identify. With information from *Bergey's Manual of Systematic Bacteriology*, which is available in the laboratory, and from your text and this manual, work out a separation scheme in outline or flowsheet

form for distinguishing genera by using the important tests (Figure 41-1). Do this before starting the exercise so that you can use your time and the media most expediently.

Because resources vary among colleges and universities, some students will isolate unknowns from natural sources for this exercise. Your instructor will explain the rules at your institution. The traditional diagnostic tests are presented in Section VIII (Exercises 35–38); and the miniaturized multitest methods for identifying enterics are here in Exercise 41, for streptococci and staphylococci, in Exercise 70.

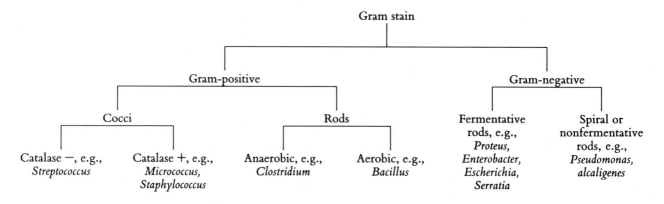

Figure 41-1   Outline of a separation scheme for identifying cultures of bacteria.

Table 41-1 Tests Used for Identifying Various Groups of Bacteria

Test	Coccus Forms[a]			Short Rod Forms						Long Rod Forms	
	Staphylococcus	Streptococcus	Micrococcus	E. coli	E. aerogenes	Proteus	Alcaligenes[d]	Serratia	Pseudomonas	Bacillus[c]	Clostridium[e]
Temperature of incubation	37°C	37°C	30°C	37°C	37°C	30°C	20°C	30°C[b]	30°C	30°C	37°C
Gram reaction and morphology (Ex. 13)	X	X	X	X	X	X	X	X	X	X	X
Motility test	X	X	X	X	X	X	X	X	X	X	X
Carbohydrate fermentations (Ex. 35)	X	X	X	X	X	X	X	X	X	X	X
Gelatin hydrolysis (Ex. 36)	X	X	X	X	X	X	X	X	X	X	X
Litmus milk reaction (Ex. 36)	X	X	X	X	X	X	X	X	X	X	X
Catalase test (Ex. 38)	X	X	X	X	X	X	X	X	X	X	X
Oxygen requirement (Ex. 28)	X	X	X	X	X	X	X	X	X	X	X
Growth in nutrient broth (Ex. 7)	X	X	X	X	X	X	X	X	X	X	X
Colony morphology (Ex. 9)	X	X	X	X	X	X	X	X	X	X	X
Growth in yeast-extract broth[a]	X	X	X								
Indole test (Ex. 36)				X	X						
MR–VP test (Ex. 35)				X	X						
Citrate utilization test (Ex. 35)				X	X						
Urease test (Ex. 37)						X					
Mannitol-agar slope								X			
Oxidase test (Ex. 38)									X		
Endospore stain[c] (Ex. 15)										X	X
Starch hydrolysis (Ex. 35)										X	
Casein hydrolysis (Ex. 36)										X	
Growth in thioglycollate broth (Ex. 29)											X

[a]Although many common bacteria grow satisfactorily in beef-extract media (such as nutrient broth or agar), the streptococci and staphylococci require a richer medium for good growth: beef infusion or yeast-extract broths.

[b]For pigment formation.

[c]If no endospores are apparent in the preparation that you use, inoculate a manganese-agar slope with your organism and perform your endospore stain from this growth. In general, endospores are produced better on agar surfaces than in broth media and are stimulated by the inclusion of as little as 2 ppm manganese in the media.

[d]A. viscolactis is notable for its ability to cause a ropiness in milk. Thrust the inoculating loop into the surface of the litmus milk culture and note a threadlike slime that adheres to the loop on withdrawal.

[e]Boil all liquid media for 10 minutes and allow to cool before inoculating with Clostridium.

Table 41-2 Differentiation of Enterobacteriaceae by Biochemical Tests

Test	Escherichia coli	Shigella sonnei	Other Shigella	Edwardsiella tarda	Typical Salmonella	Salmonella typhi	Arizona	Citrobacter freundii	Citrobacter diversus	Citrobacter amalonaticus	Klebsiella pneumoniae	Klebsiella oxytoca	Enterobacter cloacae	Enterobacter aerogenes	Enterobacter agglomerans	Enterobacter sakazakii	Enterobacter gergoviae	Hafnia alvei	Serratia marcescens	Serratia liquefaciens	Serratia rubidaea	Proteus vulgaris	Proteus mirabilis	Morganella morganii	Providencia rettgeri	Providencia alcalifaciens	Providencia stuartii	Yersinia enterocolitica	Yersinia pseudotuberculosis	Yersinia pestis
Indole	+	-	V	+	-	-	-	-	+	+	-	+	-	-	V	V	-	-	-	-	-	+	-	+	+	+	+	V	-	-
Methyl Red	+	+	+	+	+	+	+	+	+	+	V	V	-	-	V	V	V	V	V	V	V	+	+	+	+	+	+	+	+	+
Voges-Proskauer	-	-	-	-	-	-	-	-	-	-	+	+	+	+	V	+	+	V	+	V	+	-	V	-	-	-	-	V	-	-
Simmons' Citrate	-	-	-	-	V	-	+	+	+	-	+	+	+	+	V	+	+	V	+	+	(V)	V	(V)	-	+	+	+	-	-	-
Hydrogen Sulfide (TSI)	-	-	-	+	+	$+^w$	+	+	-	-	-	-	-	-	-	-	-	-	-	-	-	+	+	-	-	-	-	-	-	-
Urea	-	-	-	-	-	-	-	V^w	V^w	V	+	+	V^w	+	V^w	-	+	-	V^w	V^w	V^w	+	V	+	+	-	V	+	+	V
KCN	-	-	-	-	-	-	-	+	-	+	+	+	+	+	V	+	-	+	+	+	V	+	+	+	+	+	+	-	-	-
Motility	V	-	-	+	+	+	+	+	+	+	-	-	V	+	V	+	+	+	+	+	V	+	+	V	+	+	V	-37C/+22C	-37C/+22C	-37C/+22C
Gelatin (22°C)	-	-	-	-	-	-	(+)	-	-	-	-	-	V	-	V	-	-	-	(V)	+	(V)	+	-	-	-	-	-	-	-	-
Lysine Decarboxylase	V	-	-	+	+	+	+	-	-	-	+	+	-	+	-	-	(V)	+	+	(V)	(V)	-	-	-	-	-	-	+	-	-
Arginine Dihydrolase	V	-	V	-	(V)	-	(V)	V	(V)	+	-	-	+	-	-	+	-	V	-	-	-	-	-	-	-	-	-	-	-	-
Ornithine Decarboxylase	V	+	-	+	+	-	+	V	+	+	-	-	+	+	-	+	+	+	+	+	-	-	+	+	-	-	-	+	-	-
Phenylalanine Deaminase	-	-	-	-	-	-	-	-	-	-	-	-	-	-	V	V	-	-	-	-	-	+	+	+	+	+	+	-	-	-

Malonate	–	–	–	–	+	+	V	+	–	+	+	V	V	V	V	–	–	–	–	–	V	–	–	–	–
Gas from D-Glucose	–	–	–	+	+	+	+	+	+	+	+	+	V	V	+	+	V	V	V	+	V	–	–	–	–
Lactose	‒	–	–	–	V	+	(V)	V	V	+	V	+	V	–	–	+	–	–	–	–	–	–	–	–	–
Sucrose	V	‒	–	–	V	+	V	V	+	+	V	V	V	V	V	V	+	–	V	V	V	+	+	–	–
D-Mannitol	+	+	V	–	+	+	+	+	+	+	+	+	+	+	+	+	–	–	+	+	+	+	+	>+	–
Dulcitol	V	–	Vc	–	–	V	V	–	–	V	V	–	–	–	–	–	–	–	–	–	–	–	–	–	–
Salicin	V	V	–	–	V	+	(V)	(V)	+	+	+	+	V	V	V	V	–	–	–	V	–	V	+	+	+
Adonitol	–	–	–	–	–	V	V	–	V	V	–	–	–	–	+	+	–	–	–	–	–	–	–	–	–
i (meso) Inositol	–	–	V	–	–	+	+	V	V	–	V	V	–	+	+	+	–	–	+	(V)	V	–	(V)	V	V
D-Sorbitol	V	–	+	+	+	+	+	+	+	–	+	+	–	–	–	–	–	–	+	+	+	–	+	V	V
L-Arabinose	+	V	+	–	+	+	+	+	+	+	+	+	V	+	–	–	–	–	–	–	+	(V)	+	+	+
Raffinose	V	–	–	–	–	V	–	V	V	V	+	V	–	–	V	–	–	–	–	–	–	–	–	–	–
L-Rhamnose	V	(+)	V	–	+	+	+	+	+	+	+	+	+	(V)	+	+	–	V	V	+	+	+	+	+	+

+ = 90% or more positive within 48 h
– = less than 10% positive within 48 h
V = 10% to 89.9% positive within 48 h
(+) = 90% or more positive between 3 and 7 days
(V) = more than 50% positive within 48 h, and more than 90% positive in 3 to 7 days.
w = weak reaction
[a]Most *S. sonnei* strains are delayed positive in reaction for lactose (88%) and sucrose (85%).
[b]Some bioserotypes of *S. flexneri* produce gas from glucose.
[c]A few serotypes including *S. cholerae-suis*, *S. paratyphi* A, and *S. pullorum* do not ferment dulcitol within 48 h, *S. cholerae-suis* does not ferment arabinose.
This chart is designed to be a brief guide to the reactions of the more clinically important species of *Enterobacteriaceae*. Only 26 of the 60 or more tests used to distinguish between species are listed. Specific biotypes (H₂S+, *E. coli*, lactose[a] and raffinose[a] *Y. enterocolitica*, etc.), fastidious strains and atypical strains are not addressed. For a more sophisticated treatment of these and other species of *Enterobacteriaceae* the reader should consult special publications that give the above information and percentages.
Courtesy DIFCO Laboratories

PROCEDURE

1. If you received a pure culture, start at Step 4. If you were given a mixture of organisms by your instructor, the first step is to streak your culture on nutrient agar, EMB agar, or other media recommended to you, using the technique you learned in Exercise 9. Incubate your plates at 37°C for 18 hours unless you are instructed to do otherwise.

2. Make a gram stain of your mixture (Exercise 13) and study it in your microscope. This will let you know the types of bacteria in your unknown and aid you in the selection of colonies for further study.

3. If your colonies are clearly separated, try to recognize the colony types. Pick one colony of each type and restreak it on fresh medium. Pick other colonies of each type and make smears on microscope slides for staining.

4. Make gram stains of each colony type. Observe the gram characteristics and cell morphology of your unknown.

5. Make two transfers, one slope and one broth, of each of your unknown types. The slope will serve as your "stock" culture; the broth will be your working culture. The stock culture should not be used to inoculate various media but should be held as a reserve from which you can secure new transplants. After it has grown, the stock culture is best stored in the refrigerator or in your desk but not in the incubators. It should be transferred only as necessary (once a week in this exercise).

6. Prepare shake cultures to determine oxygen relationships.

7. From the information on gram reaction, morphology, motility, and oxygen relationships, the possible identities of your unknown cultures can be greatly narrowed. Having examined this evidence, you must determine which additional media and tests should be used for identification (see Tables 41-1, 41-2, and 41-3; Exercises 35–38). Plan your inoculations far enough in advance to allow the proper incubation period for each test. For example, proteolysis of milk requires incubation for several days before the reaction is apparent. Thus, you would not be likely to obtain the proper results if you were to inoculate litmus milk with an unknown culture 24 hours before your results were due.

8. Carry out the physiological tests necessary for the generic or specific identification of each of your unknowns.

9. Record the results on your report sheets.

Table 41-3 Interpreting Enteric Reactions

Medium	Observations	Conclusions
TSI (triple sugar iron) slant Glucose, lactose, sucrose	1. Slant red, butt yellow	Only glucose fermented
	2. Slant and butt yellow	Glucose and lactose and/or sucrose fermented
pH indicator is phenol red	3. Bubbles, cracks, agar pushed up	Gas produced due to formic hydrogenlyase
	4. Black precipitate	H_2S production
Phenylalamine slant, reagent: $FeCl_2$	Surface of slant turns dark green after flooding with $FeCl_2$	Phenylalanine deaminase produced
Simmons' citrate slant pH indicator is bromthymol blue	Slant turns from green to dark blue	Citrate is metabolized
SIM (sulfide, indole, motility)	1. Growth away from stab	Motility
	2. Black precipitate	H_2S production
Reagent is Kovacs'	3. Red color after addition of 4–6 drops Kovacs'	Indole is produced from tryptophan
MIO (motility, indole, ornithine) pH indicator is bromcresol purple	1. Growth away from stab	Motility
	2. Yellow throughout or at bottom	No decarboxylation
	3. Faded purple throughout	Glucose fermented
Reagent is Kovacs' (seal with vaspar!)	4. Red color after 4–6 drops Kovacs' added (remove vaspar before adding reagent)	Ornithine decarboxylated Indole produced

(continued)

Table 41-3 *(continued)*

Medium	Observations	Conclusions
Decarboxylase broth		
with lysine	Purple	Lysine decarboxylated
without lysine	Yellow	Glucose fermented
pH indicator is bromcresol purple	NOTE: If tube with lysine is the same color as tube without lysine, the test is negative for decarboxylation.	
Seal with vaspar		
MR–VP (methyl red– Voges-Proskauer)	Pink to red immediately when 5 drops methyl red are added to 3 ml culture	Mixed acid fermentation
MR reagent–methyl red		
VP—reagents—napthol, 40% KOH–creatine	Reddish color after adding 4 drops of each reagent to 3 ml culture. Shake well, warm gently, wait up to 30 min for reaction.	Acetylmethylcarbinol present– butanediol fermentation
Urea broth pH indicator is phenol red	Bright pink color	Urease is produced
BCP carbohydrate broth	Yellow	Acid produced
Glucose and Lactose Durham tube	Gas in Durham tube	Formic hydrogenlyase produced
	Yellow	Acid produced, lactose fermented
pH indicator is bromcresol purple		

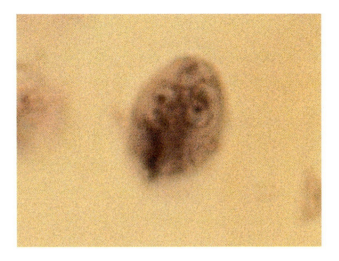

Color Plate 17 Hematoxlin stain of *Giardia muris* from the small intestine of a rat. (Photo by J. J. Lee)

Color Plate 18 Arranged slide showing various shapes in diatoms. (Photo by J. J. Lee)

Color Plate 19 Petri dish cultures of two common molds: *Aspergillus* (left) and *Rhizopus* (right). (Photos by J. J. Lee)

Color Plate 20 *Rhizopus* sporangium. (Photo by J. J. Lee)

Color Plate 21 Zygosporangium of *Rhizopus*. (Photo by J. J. Lee)

Color Plate 13 *Biddulphia,* a diatom, photographed with the aid of interference microscopy. (Photo by J. J. Lee)

Color Plate 14 *Lichmophora,* a diatom, photographed with the aid of interference microscopy. (Photo by J. J. Lee)

Color Plate 15 *Isotricha* (left) and *Ophryoscolex* (right), ciliated protists from sheep rumen. (Photos by J. J. Lee)

Color Plate 16 Hematoxlin stain of *Zelleriella* from the large intestine of a frog. (Photo by J. J. Lee)

Color Plate 7   Acetoorcein stain of *Spirostomum,* a ciliated protist with a ribbon–like macronucleus. (Photo by J. J. Lee)

Color Plate 8   Acetoorcein stain of a ciliated protist, *Stentor.* (Photo by J. J. Lee)

Color Plate 9   *Lithodesmium,* a diatom, photographed with the aid of interference microscopy. (Photo by J. J. Lee)

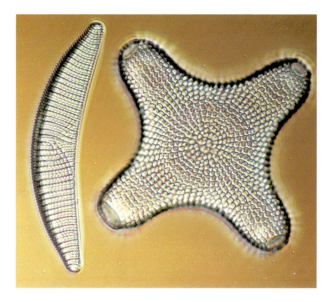

Color Plate 10   *Epithemia* (left) and *Triceratium* (right), diatoms, photographed with the aid of interference microscopy. (Photo by J. J. Lee)

Color Plate 11   *Actinocyclus* (left) and *Epithemia* (right), diatoms, photographed with the aid of interference microscopy. (Photo by J. J. Lee)

Color Plate 12   *Melosira,* a diatom, photographed with the aid of interference microscopy. (Photo by J. J. Lee)

V.P.			**Voges-Proskauer** Acetylmethylcarbinol (acetoin) is an intermediate in the production of butylene glycol from glucose fermentation. The presence of acetoin is indicated by the development of a red color within 20 minutes. Most positive reactions are evident within 10 minutes.
DUL-PA			**Dulcitol** Bacterial fermentation of dulcitol, which results in the formation of acidic end products, is indicated by a change in color of the indicator present in the medium from green (alkaline) to yellow or pale yellow (acidic).
		A★ B★	**Phenylalanine deaminase** This test detects the formation of pyruvic acid from the deamination of phenylalanine. The pyruvic acid formed reacts with a ferric salt in the medium to produce a characteristic black to smoky gray color.
UREA			**Urea** The production of urease by some bacteria hydrolyzes urea in this medium to produce ammonia, which causes a shift in pH from yellow (acidic) to reddish-purple (alkaline). This test is strongly positive for *Proteus* in 6 hours and weakly positive for *Klebsiella* and some *Enterobacter* species in 24 hours.
CIT		A★ B★	**Citrate** Organisms that are able to utilize the citrate in this medium as their sole source of carbon produce alkaline metabolites which change the color of the indicator from green (acidic) to deep blue (alkaline). Any degree of blue should be considered positive.

★ Certain microorganisms will not always produce the ideal "strong" positive A color. Reactions similar to B should also be considered positive.

Color Plate 6 The API 20E strip: *Proteus vulgaris* (top) and *Escherichia coli* (bottom). (Photo by J. J. Lee)

Color Plate 5 Biochemical reactions of Enterotube II. (Courtesy of Roche Diagnostics, Nutley, N.J.)

Symbol	Uninoculated color	Reacted color	Type of reaction
GLU–GAS			**Glucose (GLU)** The end products of bacterial fermentation of glucose are either acid or acid and gas. The shift in pH due to the production of acid is indicated by a color change from red (alkaline) to yellow (acidic). Any degree of yellow should be interpreted as a positive reaction; orange should be considered negative.
			Gas production (GAS) Complete separation of the wax overlay from the surface of the glucose medium occurs when gas is produced. The amount of separation between the medium and overlay will vary with the strain of bacteria.
LYS			**Lysine decarboxylase** Bacterial decarboxylation of lysine, which results in the formation of the alkaline end product cadáverine, is indicated by a change in the color of the indicator from pale yellow (acidic) to purple (alkaline). Any degree of purple should be interpreted as a positive reaction. The medium remains yellow if decarboxylation of lysine does not occur.
ORN			**Ornithine decarboxylase** Bacterial decarboxylation of ornithine causes the alkaline end product putrescine to be produced. The acidic (yellow) nature of the medium is converted to purple as alkalinity occurs. Any degree of purple should be interpreted as a positive reaction. The medium remains yellow if decarboxylation of ornithine does not occur.
H_2S/IND			**H_2S production** Hydrogen sulfide, liberated by bacteria that reduce sulfur-containing compounds such as peptones and sodium thiosulfate, reacts with the iron salts in the medium to form a black precipitate of ferric sulfide usually along the line of inoculation. Some *Proteus* and *Providencia* strains may produce a diffuse brown coloration in this medium, which should not be confused with true H_2S production.
			Indole formation The production of indole from the metabolism of tryptophan by the bacterial enzyme tryptophanase is detected by the development of a pink to red color after the addition of Kovac's reagent.
ADON			**Adonitol** Bacterial fermentation of adonitol, which results in the formation of acidic end products, is indicated by a change in color of the indicator present in the medium from red (alkaline) to yellow (acidic). Any sign of yellow should be interpreted as a positive reaction; orange should be considered negative.
LAC			**Lactose** Bacterial fermentation of lactose, which results in the formation of acidic end products, is indicated by a change in color of the indicator present in the medium from red (alkaline) to yellow (acidic). Any sign of yellow should be interpreted as a positive reaction; orange should be considered negative.
ARAB			**Arabinose** Bacterial fermentation of arabinose, which results in the formation of acidic end products, is indicated by a change in color from red (alkaline) to yellow (acidic). Any sign of yellow should be interpreted as a positive reaction; orange should be considered negative.
SORB			**Sorbitol** Bacterial fermentation of sorbitol, which results in the formation of acidic end products, is indicated by a change in color from red (alkaline) to yellow (acidic). Any sign of yellow should be interpreted as a positive reaction; orange should be considered negative.

Color Plate 2 Endospore stain in *Bacillus subtilus.* (Photo by J. J. Lee)

Color Plate 1 Acid-fast stained *Mycobacterium tuberculosis* (arrow) in sputum. (Photo by J. J. Lee)

Color Plate 3 *Peritrichous* flagella stained by Leifson's method. (Photo by J. J. Lee)

Color Plate 4 Enterotube II: uninoculated (top), *Proteus vulgaris* (middle), and *Escherichia coli* (bottom). (Photo by J. J. Lee)

Color Plate 22   Conidiophores of *Penicillium* (left) and *Aspergillus* (right). (Photos by J. J. Lee)

Color Plate 23   *Candida albicans* grown on cornmeal agar to stimulate chlamydospore production. (Photo by J. J. Lee)

Color Plate 24   *Streptococcus pneumoniae* (arrow) from a human sputum sample. (Photo by J. J. Lee)

Color Plate 25   Two typical primary throat culture isolations on blood agar. (Photos by J. J. Lee)

| (α) Alpha | (β) Beta | (γ) Gamma |

Color Plate 26   Typical reactions of bacteria on blood agar. (Photos by J. J. Lee)

Color Plate 27   API Rapid STREP. (Courtesy API)

Identifying Unknown
Bacterial Cultures: Maximethods

Name _____

Desk No. _____

Culture No. Received _____

Genus _____

PART A: MORPHOLOGY

VEGETATIVE CELLS

Form	spheres, short rods, long rods, filaments, commas, spirals
Arrangement	singles, pairs, chains, fours, cubical packets, clusters
Capsules	absent, present
Staining reaction	gram- _____
Motility	present, absent
Spores	present, absent
Location	central, subterminal, terminal

PART B: CULTURAL CHARACTERISTICS

COLONY

Medium: _____ Age: _____

Growth	slow, rapid, moderate
Form	punctiform, circular, rhizoid, irregular
Surface	smooth, rough, dry, moist, dull, glistening
Elevation	flat, raised
Edge	smooth, wavy, filamentous

SLANT

Medium: _____

Growth	scanty, moderate, abundant
Form	threadlike, beaded, spreading, rootlike
Elevation	flat, raised
Optical characters	opaque, translucent, iridescent
Color	_____ water-soluble, water-insoluble
Consistency	butyrous, viscid, brittle

BROTH

Medium: _____

Surface growth	ring, pellicle, none
Clouding	slight, heavy, none
Amount of sediment	abundant, scanty, none
Type of sediment	flaky, granular, viscid on agitation

OXYGEN RELATION

_____ : in _____ (medium)

TEMPERATURE OPTIMUM: _____ °C

PART C: PHYSIOLOGY

GELATIN STAB

    Liquefaction   none, slow, moderate, complete

LITMUS MILK

    Reduction   slight, moderate, complete, none

    Incubation time: _____

FERMENTATIONS

    Glucose   acid, gas, negative

    Incubation time: _____

    Lactose   acid, gas, negative

    Incubation time: _____

    Sucrose   acid, gas, negative

    Incubation time: _____

PART D: ADDITIONAL DATA

ENDO OR EOSIN METHYLENE BLUE AGAR
    Growth   present, absent
    Color   present, absent
    Metallic sheen   present, absent

SPECIFIC TESTS *(indicate positive or negative)*

Indole
Methyl red
Acetylmethylcarbinol (V-P)
Oxidase
Lipid hydrolysis
Lysine decarboxylation
Ornithine decarboxylation
Phenylalanine deamination
Catalase
Starch hydrolysis
Casein hydrolysis
Urease
Citrate utilization

Identifying of Unknown
Bacterial Cultures: Maximethods

Name _____

Desk No. _____

Culture No. Received _____

Genus _____

PART A: MORPHOLOGY

VEGETATIVE CELLS

Form	spheres, short rods, long rods, filaments, commas, spirals
Arrangement	singles, pairs, chains, fours, cubical packets, clusters
Capsules	absent, present
Staining reaction	gram- _____
Motility	present, absent
Spores	present, absent
Location	central, subterminal, terminal

PART B: CULTURAL CHARACTERISTICS

COLONY

Medium: _____ Age: _____

Growth	slow, rapid, moderate
Form	punctiform, circular, rhizoid, irregular
Surface	smooth, rough, dry, moist, dull, glistening
Elevation	flat, raised
Edge	smooth, wavy, filamentous

SLANT

Medium: _____

Growth	scanty, moderate, abundant
Form	threadlike, beaded, spreading, rootlike
Elevation	flat, raised
Optical characters	opaque, translucent, iridescent
Color	_____ water-soluble, water-insoluble
Consistency	butyrous, viscid, brittle

BROTH

Medium: _____

Surface growth	ring, pellicle, none
Clouding	slight, heavy, none
Amount of sediment	abundant, scanty, none
Type of sediment	flaky, granular, viscid on agitation

OXYGEN RELATION

_____: in _____ (medium)

TEMPERATURE OPTIMUM: _____°C

PART C: PHYSIOLOGY

GELATIN STAB

    Liquefaction   none, slow, moderate, complete

LITMUS MILK

    Reduction   slight, moderate, complete, none

    Incubation time: _____

FERMENTATIONS

    Glucose   acid, gas, negative

    Incubation time: _____

    Lactose   acid, gas, negative

    Incubation time: _____

    Sucrose   acid, gas, negative

    Incubation time: _____

PART D: ADDITIONAL DATA

ENDO OR EOSIN METHYLENE BLUE AGAR
        Growth   present, absent
         Color   present, absent
  Metallic sheen   present, absent

SPECIFIC TESTS *(indicate positive or negative)*

Indole
Methyl red
Acetylmethylcarbinol (V-P)
Oxidase
Lipid hydrolysis
Lysine decarboxylation
Ornithine decarboxylation
Phenylalanine deamination
Catalase
Starch hydrolysis
Casein hydrolysis
Urease
Citrate utilization

MINIATURIZED MULTITEST METHODS FOR IDENTIFYING ENTERICS

A. ENTEROTUBE II*

The family Enterobacteriaceae is composed of a large number of closely related small-gram negative rods. Their family name and their designation as enterics are based on the fact that the natural habitat of most of them is the intestine of man and animals. Many are pathogens that cause a variety of gastrointestinal diseases as well as nosocomial infections. The rapid and accurate identification of these pathogens is essential for the treatment of the diseases they cause.

Traditionally, the identity of an enteric species is determined by a series of sugar fermentations and biochemical tests performed in tubed media as illustrated in Section VIII. Because these tests are costly and time-consuming, they are being replaced by miniaturized microbial identification systems. These miniaturized systems employ multicompartment–multitest kits that require less media and provide rapid results that are adaptable to computer data processing. In this exercise you will have the opportunity to use one of the many commercial kits to identify an unknown enteric culture.

The **Enterotube II** consists of a plastic tube with 12 compartments each of which contains a different medium (Figure 41-2 and Color Plate 4). An inoculating wire passes through the various media and protrudes at both ends of the tube. For inoculation, one end of the protruding wire is touched to the center of a colony of the test microbe to be identified and then is withdrawn through the compartments, inoculating all the test media. It is then reinserted into four compartments of the Enterotube to maintain reducing or anaerobic conditions.

The enteric culture can then be identified by conventional patterns of positive and negative fermentation and biochemical reactions, or by a computerized code system. In the latter, the combination of biochemical reactions is reduced to an identification (ID) value of a 5- or 7-digit profile number using binary mathematics. These numbers are then used with computer printouts to achieve the best fit of biochemical tests and to make an identification. This computerized best-fit matching takes into account normal biological variations that lead to mistakes early in the identification procedure.

The computer-assisted identification has additional advantages in that all possible combinations of biochemical reactions are taken into consideration, and a probability score is available for identifying atypical organisms.

There are some disadvantages to minitest systems: The accuracy of some tests is questionable; interpretation of reactions can be difficult; and sometimes it is necessary to perform other biochemical, serological, or phage identification tests. This experiment is designed to enable you to recognize the significant tests that define the members of the Enterobacteriaceae.

You are provided with plate cultures of two enteric bacteria as unknowns. *Caution: Take care in working with these unknowns because some are pathogens.*

PROCEDURE

1. You are provided with two Enterotubes and a computer coding and identification booklet.

*Further information on the Enterotube II may be obtained by writing Roche Diagnostic's Division of Hoffmann-La Roche, Inc. Nutley, NJ 07110.

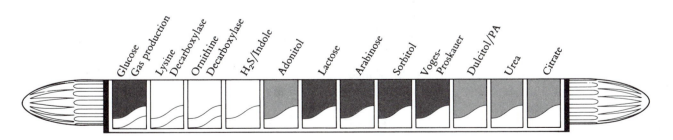

Figure 41-2 The Enterotube II.

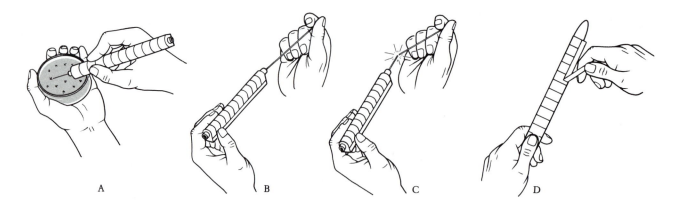

Figure 41-3 Inoculation of the Enterotube II. *A*: Pick a well-isolated colony with the Enterotube II inoculating needle. *B*: Inoculate the compartments by withdrawing the needle through the tube. *C*: Break the reinserted needle at the black ring on its surface. *D*: Strip off the blue tape.

2. Remove the caps from both ends of one Enterotube. *Do not flame needle.*

3. Pick a well-isolated colony directly with the tip of the Enterotube inoculating needle (Figure 41-3A).

4. Inoculate the Enterotube by twisting the wire and then withdrawing the needle through all 12 compartments using a twisting-turning motion (Figure 41-3B).

5. Reinsert the needle into the Enterotube using a turning motion through the first three compartments (glucose, lysine, and ornithine) into the H₂S/indole compartment. Break off the needle by bending it at the notch that is now at the mouth of the Enterotube and discard it (Figure 41-3C). The iron wire insures the necessary anaerobic conditions for glucose fermentation and the decarboxylation of lysine and ornithine in these three compartments.

6. Replace the caps on the tube.

7. Strip off the blue tape to expose the holes in the plastic tube and provide aerobic conditions in the adonitol, lactose, arabinose, sorbitol, Voges–Proskauer, dulcitol/phenylalanine, urea, and citrate compartments (Figure 41-3D).

8. Slide a clear plastic band over the glucose compartment to contain any wax that may escape due to abundant gas production by some bacteria.

9. Repeat this inoculation procedure (Steps 2–8) with the second Enterotube and the other culture to be identified.

10. Incubate both Enterotubes at 35–37°C for 24 hours with the tubes lying on their flat surfaces.

11. Interpret and record all media reactions (except indole) by comparing the reactions in each tube with an uninoculated Enterotube and the chart of positive reaction colors provided (Color Plate 5).

12. Perform the indole test by placing the Enterotube in a rack with the glucose compartment pointing downward and, using a needle and syringe, add 2 or 3 drops of Kovacs reagent through the plastic film of the H₂S/indole compartment (Figure 41-4). Allow the reagent to contact the agar surface and read after 1 minute.

13. Perform the Voges–Proskauer test by placing the Enterotube in a rack with the glucose compartment pointing downward and add 2

Figure 41-4 Performing the indole test.

drops of 20% potassium hydroxide containing 0.3% creatine and 3 drops of 5% alpha-naphthol solution. The development of a red color within 20 minutes indicates a positive test.

14. On your report sheet indicate each positive reaction by circling the number appearing below the appropriate compartment of the Enterotube II outline (Figure 41-5A). Add the circled numbers in each bracket section and enter the sum in the space provided below the arrow on the report sheet (Figure 41-5B).

15. Read the five numbers in these spaces across as a 5-digit number. To obtain the identity of the genus and/or species of the unknown bacteria, find this 5-digit number in the computer coding manual in the column entitled ID Value (Figure 41-5C). If more than one organism is listed, further confirmatory tests are needed for a positive identification. These confirmatory tests are listed in the right-hand column of the identification booklet.

B. THE API 20E SYSTEM*

The API 20E System is a standardized, miniaturized version of 23 conventional biochemical tests for the identification of Enterobacteriaceae and other gram-negative bacteria. It can be used for same-day (Figure 41-6) and 18–24 hour identification of Enterobacteriaceae as well as 18–24 or 36–48 hour identification (Figure 41-7) of other gram-negative bacteria. As with other biochemical identification systems, the results should be considered presumptive, and the identification of *Salmonella, Salmonella* subgroup 3, *Shigella, E. coli* A–D, and *Vibrio cholerae* from gastroenteritis should be confirmed serologically. The system consists of microtubules containing dehydrated substrates that have a reasonable shelf life (18 months) if stored in a dark refrigerator (2–8°C).

PROCEDURE

Use the following procedures, which are supplied by the manufacturers of the API 20E System.

QUESTIONS

1. What is a nosocomial infection?
2. What other test might be run to identify nonenteric bacteria by these miniaturized methods?

---

*Further information on the API 20E System can be obtained from the manufacturer, Analytab Products, 200 Express St., Plainview, NY 11803.

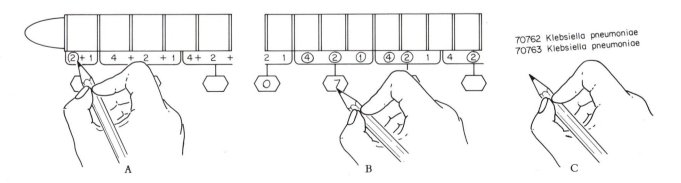

Figure 41-5 Evaluation of the Enterotube II: results and identification of species. *A*: Indicate each positive reaction by circling the number below the appropriate compartment. *B*: Add the circled numbers within each bracket section and enter this sum into the space below. *C*: Find the 5-digit number in the coding manual to identify the culture.

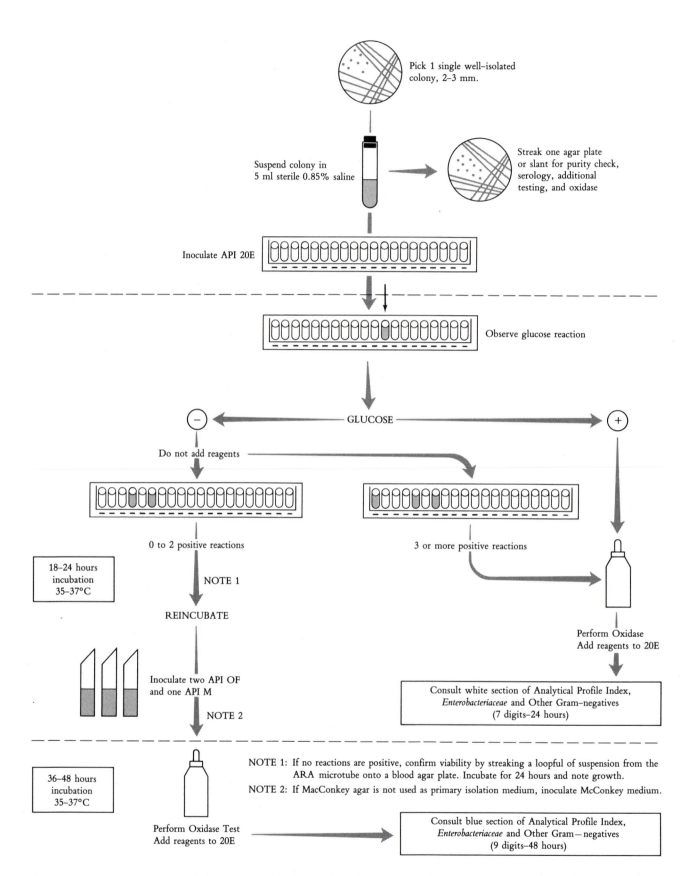

Figure 41-6 Flow chart of the recommended procedure for the identification of the Enterobacteriaceae and other gram-negative bacteria by the API 20E test battery. Same-day protocol.

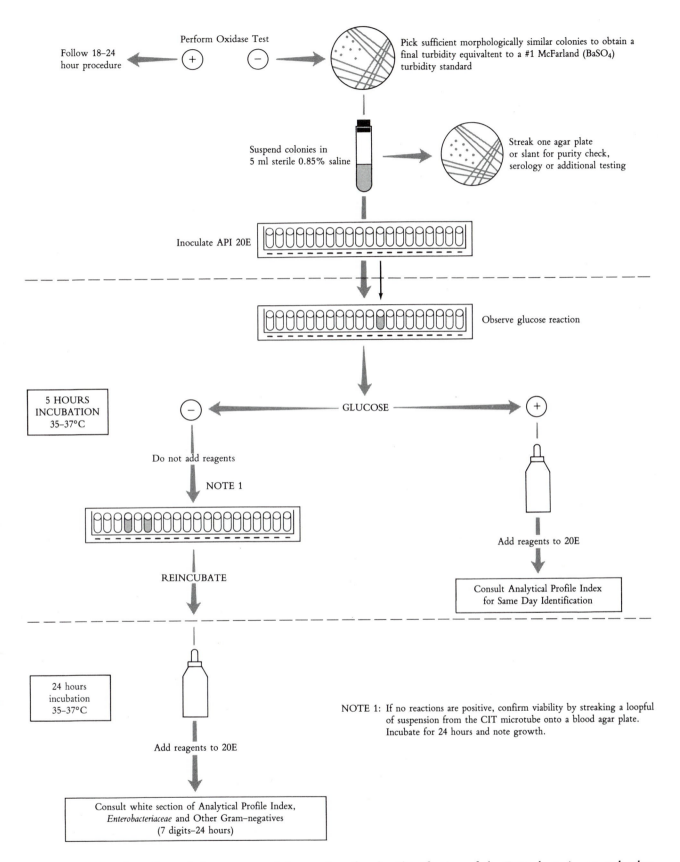

Perform Oxidase Test

Follow 18–24 hour procedure ← (+) (−) → Pick sufficient morphologically similar colonies to obtain a final turbidity equivalent to a #1 McFarland (BaSO₄) turbidity standard

Suspend colonies in 5 ml sterile 0.85% saline

Streak one agar plate or slant for purity check, serology or additional testing

Inoculate API 20E

Observe glucose reaction

5 HOURS INCUBATION 35–37°C ← (−) GLUCOSE (+) →

Do not add reagents

NOTE 1

REINCUBATE

Add reagents to 20E

Consult Analytical Profile Index for Same Day Identification

24 hours incubation 35–37°C

NOTE 1: If no reactions are positive, confirm viability by streaking a loopful of suspension from the CIT microtube onto a blood agar plate. Incubate for 24 hours and note growth.

Add reagents to 20E

Consult white section of Analytical Profile Index, *Enterobacteriaceae* and Other Gram–negatives (7 digits–24 hours)

Figure 41-7 Flow chart of the recommended procedure for the identification of the Enterobacteriaceae and other gram-negative bacteria by API 20E test battery. 18–24 hour protocol.

18–24 Hour Procedure

1. PREPARATION OF BACTERIAL SUSPENSION

    (1) Add 5 ml of 0.85% saline, pH 5.5–7.0 to a sterile test tube.
    Note: Saline containing preservatives or bacteriostatic agents should *not* be used in preparing the bacterial suspension.

    (2) Gently touch the center of a well-isolated colony (2 or 3 mm or larger in diameter) with the tip of a wooden applicator stick. Insert the applicator stick into the tube of saline and, with the tip of the stick at the base of the tube, rotate the stick in a vortex-like action. Recap the tube. Shake vigorously. The turbidity of the suspension should match McFarland No. 3 barium sulfate suspension (Figure 33-8B).

    Alternate procedure: With a flamed inoculating loop, carefully touch the center of a well-isolated colony (2 or 3 mm or larger in diameter) and thoroughly mix the inoculum with the tubed saline.

2. PREPARATION OF STRIPS

    (1) Set up an incubation tray and lid.
    (2) Identify yourself and the specimen on the elongated flap of the tray.
    (3) Dispense 5 ml of tap water into the incubation tray to provide a humid atmosphere during incubation. A plastic squeeze bottle can be used for this (Figure 33-8A).
    (4) Remove the API strips from the sealed pouch and place one strip in each incubation tray. (Figure 41-8C).

3. INOCULATION OF THE STRIPS

    The API 20E strip contains 20 microtubes (Color Plate 6), each of which consists of a tube and a cupule section (Figure 41-8D).

    (1) Remove the cap from the tube containing the bacterial suspension and insert a 5 ml Pasteur pipet.
    (2) Tilt the API 20E incubation tray and fill the tube section of the microtubes by placing the pipet tip against the side of the cupule.
    Note: The ADH, LDC, ODC, H_2S, and

URE reactions can be interpreted best if these microtubes are slightly underfilled.

    (3) Fill both the TUBE and CUPULE section of the (CIT), (VP), and (GEL) tubes.
    (4) After inoculation, completely fill the cupule section of the ADH, LDC, ODC, H_2S, and URE tubules with mineral oil (Figure 41-8E)
    (5) Using the excess bacterial suspension, inoculate an agar slant or plate (nonselective media such as nutrient agar, blood agar, or tryptic (trypticase) soy agar is suggested) as a purity check and for oxidase testing, serology, and/or additional biochemical testing. Incubate the slant or plate for 18–24 hours at 35–37°C.

4. INCUBATION OF THE STRIPS

    (1) After inoculation, place the plastic lid on the tray and incubate the strip for 18–24 hours at 35–37°C in a non-CO_2 incubator.
    (2) Weekend incubation: The biochemical reactions of the API 20E should be read after 18–24 hours incubation. If the strips cannot be read after 24 hours incubation at 35–37°C, the strips should be removed from the incubator and stored at 2–8°C (refrigerator) until the reactions can be read.

5. READING THE STRIPS

    (1) After 18 hours and before 24 hours incubation, record all reactions not requiring the addition of reagents. Interpretation of the reactions can be found in Tables 41-4 and 41-5.
    (2) If the GLU tube is negative (blue or green), do not add reagents. Consult "Recommended Procedure for the Identification of Enterobacteriaceae and Other gram-negative Bacteria (18–24 Hour)," found in Table 41-5.
    (3) If the GLU tube is positive (yellow):

        (a) Perform the oxidase test.
        A portion of the growth from the agar slant or plate, inoculated from the 20E bacterial suspension, should be rubbed onto filter paper to which a drop of oxidase reagent (1% tetramethyl-*p*-phenylenediamine dihydrochloride) has been added. The

ONPG	ADH	LDC	ODC	CIT	H₂S	URE	TDA	IND	VP		GEL	GLU	MAN
1	2	4	1	2	4	1	2	4	1		2	4	1
+	–	+	+	–	–	–	–	+	–		–	+	+

Figure 41-8 Some of the steps in the API 20E procedure: *A*: Place 5 ml of tap water into bottom of tray. *B*: Take loopfuls of organisms to make a suspension in saline that matches the McFarland No. 3 standard. *C*: Place test strip in bottom of moistened tray. *D*: Inoculate suspension into all 20 compartments; underfill ADH, ODC, H₂S, and URE; fill the tube and open cupule section of CIT, VP, and GEL with the saline cell suspension. *E*: To provide anaerobic conditions for ADH, ODC, LDC, H₂S, and URE, add sterile mineral oil to cupules. *F*: After incubation and the addition of test reagents to the appropriate compartments, record test results and total numbers to arrive at 7-digit code.

Table 41-4 API 20E Summary of Results—Same-Day Procedure

Tube	Positive	Negative	Comments
ONPG	Yellow	Colorless	(1) Any shade of yellow is a positive reaction. (2) VP tube, before the addition of reagents, can be used as a negative control.
ADH	Red or Orange	Yellow	Any shade of orange or red is a positive reaction.
LDC	Red or Orange	Yellow	Any shade of orange or red is a positive reaction.
ODC	Red or Orange	Yellow	Any shade of orange or red is a positive reaction.
CIT	Turquoise or Dark Blue	Light Green or Yellow	(1) Both the tube and cupule should be filled. (2) Reaction is read in the aerobic (cupule) area. (3) Any shade of blue should be interpreted as positive.
H_2S	Black Deposit	No Black Deposit	(1) H_2S production may range from a heavy black deposit to a very thin black line around the tube bottom. Carefully examine the bottom of the tube before considering the reaction negative. (2) A "browning" of the medium is a negative reaction unless a black deposit is present. "Browning" occurs with TDA positive organisms.
URE	Red or Orange	Yellow	(1) A method of lower sensitivity has been chosen. *Proteus* and *Yersinia* routinely give positive reactions. (2) Any shade of orange or red should be interpreted as positive.
TDA	Add 1 drop 10% Ferric chloride Red Brown	Yellow	Immediate reaction
IND	Add 1 drop Kovacs Reagent Pink-Red Ring	Yellow	(1) The reaction should be read within 2 minutes after the addition of the Kovacs reagent and the results recorded. (2) After several minutes, the HCl present in Kovacs reagent may react with the plastic of the cupule resulting in a change from a negative (yellow) color to a brownish-red. This is a negative reaction.
VP	Add 1 drop of 40% Potassium hydroxide, then 1 drop of 6% Alpha-naphthol Pink-Red	Colorless	(1) Wait 10 minutes before considering the reaction negative. (2) A pale pink color (after 10 min.) should be interpreted as negative. A pale pink color which appears immediately after the addition of reagents but which turns dark pink or red after 10 min. should be interpreted as positive.
			Motility may be observed by hanging drop or wet mount preparation.

	Diffusion of the Pigment	No Diffusion	
GEL			(1) The solid gelatin particles may spread throughout the tube after inoculation. Unless diffusion occurs, the reaction is negative. (2) Any degree of diffusion is a positive reaction.
GLU	Yellow or Gray (at the bottom of tube or throughout tube)	Blue, Blue-Green or Green	**COMMENTS FOR** **Fermentation (Enterobacteriaceae, *Aeromonas*, *Vibrio*)** (1) Fermentation of the carbohydrates begins in the most anaerobic portion (bottom) of the tube. Therefore, these reactions should be read from the bottom of the tube to the top. (2) A yellow color at the bottom of the tube only indicates a weak or delayed positive reaction.
MAN INO SOR RHA SAC MEL AMY ARA	Yellow/Yellow-Green (at bottom of tube or throughout tube)	Blue, Blue-Green or Green	**ALL CARBOHYDRATES** **Oxidation (other gram-negatives)** (1) Oxidative utilization of the carbohydrates begins in the most aerobic portion (top) of the tube. Therefore, these reactions should be read from the top to the bottom of the tube. (2) A yellow color in the upper portion of the tube and a blue in the bottom of the tube indicates oxidative utilization of the sugar. This reaction should be considered positive only for non-Enterobacteriaceae gram-negative rods. This is a negative reaction for fermentative organisms such as Enterobacteriaceae. NOTE: With the 20E 5 hr., only fermentations should be interpreted as positive.
GLU	After reading GLU reaction, add 2 drops 0.8% sulfanilic acid and 2 drops 0.5% N, N dimethyl-alpha-naphthylamine		(1) Before addition of reagents, observe GLU tube (positive or negative) for bubbles. Bubbles are indicative of reduction of nitrate to the nitrogenous (N_2) state. (2) A positive reaction may take 2–3 minutes for the red color to appear. (3) Confirm a negative test by adding zinc dust or 20 mesh granular zinc. A pink-orange color after 10 minutes confirms a negative reaction. A yellow color indicates reduction of nitrate to the nitrogenous (N_2) state.
Nitrate reduction — NO_2	Red	Yellow	
N_2 gas	Bubbles; Yellow after reagents and zinc	Orange after reagents and zinc	

Table 41-5 API 20E Summary of Results— 18–24 Hour Procedure

Tube		Interpretation of Reactions		
		Positive	Negative	Comments
ONPG		Yellow	Colorless	(1) Any shade of yellow is a positive reaction. (2) VP tube, before the addition of reagents, can be used as a negative control.
ADH	Incubation 18–24 h	Red or Orange	Yellow	Orange reactions occurring at 36–48 hours should be interpreted as negative.
	36–48 h	Red	Yellow or Orange	
LDC	18–24 h	Red or Orange	Yellow	Any shade of orange within 18–24 hours is a positive reaction. At 36–48 hours, orange decarboxylase reactions should be interpreted as negative.
	36–48 h	Red	Yellow or Orange	
ODC	18–24 h	Red or Orange	Yellow	Orange reactions occurring at 36–48 hours should be interpreted as negative.
	36–48 h	Red	Yellow or Orange	
CIT		Turquoise or Dark Blue	Light Green or Yellow	(1) Both the tube and cupule should be filled. (2) Reaction is read in the aerobic (cupule) area.
H_2S		Black Deposit	No Black Deposit	(1) H_2S production may range from a heavy black deposit to a very thin black line around the tube bottom. Carefully examine the bottom of the tube before considering the reaction negative. (2) A "browning" of the medium is a negative reaction unless a black deposit is present. "Browning" occurs with TDA positive organisms.
URE	18–24 h	Red or Orange	Yellow	A method of lower sensitivity has been chosen. *Klebsiella*, *Proteus*, and *Yersinia* routinely give positive reactions.
	38–48 h	Red	Yellow or Orange	
TDA	Add 1 drop 10% Ferric chloride			(1) Immediate reaction. (2) Indole-positive organisms may produce a golden orange color due to indole production. This is a negative reaction.
		Brown-Red	Yellow	
IND	Add 1 drop Kovacs Reagent			(1) The reaction should be read within 2 minutes after the addition of Kovacs reagent and the results recorded. (2) After several minutes, the HCl present in Kovacs reagent may react with the plastic of the cupule resulting in a change from a negative (yellow) color to a brownish-red. This is a negative reaction.
		Red Ring	Yellow	

Test	Reagent / Procedure	Positive	Negative	Interpretation
VP	Add 1 drop of 40% Potassium hydroxide, then 1 drop of 6% alpha-naphthol	Red	Colorless	(1) Wait 10 minutes before considering the reaction negative. (2) A pale pink color (after 10 min.) should be interpreted as negative. A pale pink color which appears immediately after the addition of reagents but which turns dark pink or red after 10 min. should be interpreted as positive.
				Motility may be observed by hanging drop or wet mount preparation.
GEL		Diffusion of the Pigment	No Diffusion	(1) The solid gelatin particles may spread throughout the tube after inoculation. Unless diffusion occurs, the reaction is negative. (2) Any degree of diffusion is a positive reaction.
GLU		Yellow or Gray	Blue, Blue-Green	**Fermentation** (Enterobacteriaceae, *Aeromonas*, *Vibrio*) (1) Fermentation of the carbohydrates begins in the most anaerobic portion (bottom) of the tube. Therefore, these reactions should be read from the bottom of the tube to the top. (2) A yellow color at the bottom of the tube only indicates a weak or delayed positive reaction.
MAN INO SOR RHA SAC MEL AMY ARA		Yellow	Blue or Blue-Green	**Oxidation** (other gram-negatives) (1) Oxidative utilization of the carbohydrates begins in the most aerobic portion (top) of the tube. Therefore, these reactions should be read from the top to the bottom of the tube. (2) A yellow color in the upper portion of the tube and a blue color in the bottom of the tube indicates oxidative utilization of the sugar. This reaction should be considered positive only for non-Enterobacteriaceae gram-negative rods. This is a negative reaction for fermentative organisms such as Enterobacteriaceae.
GLU	After reading GLU reaction, add 2 drops 0.8% sulfanilic acid and 2 drops 0.5% N,N-dimethyl-alpha-naphthylamine			(1) Before addition of reagents, observe GLU tube (positive or negative) for bubbles. Bubbles are indicative of reduction of nitrate to the nitrogenous (N_2) state. (2) A positive reaction may take 2–3 minutes for the red color to appear. (3) Confirm a negative test by adding zinc dust or 20 mesh granular zinc. A pink-orange color after 10 minutes confirms a negative reaction. A yellow color indicates reduction of nitrate to the nitrogenous (N_2) state.
Nitrate reduction	NO_2 / N_2 gas	Red; Bubbles; Yellow after reagents and zinc	Yellow; Orange after reagents and zinc	
MAN INO SOR Catalase	After reading carbohydrate reaction, add 1 drop 1.5% H_2O_2	Bubbles	No bubbles	(1) Bubbles may take 1–2 minutes to appear. (2) Best results will be obtained if the test is run in tubes which have no gas from fermentation.

area where the growth has been added will turn dark purple within 30 seconds if the reaction is positive and will be colorless or light purple if negative.

Note: (a) Nichrome wire loops should NOT be used in performing the oxidase test. Nicrome wire can cause a false positive reaction. Use a toothpick.

(b) The oxidase test should NOT be performed using bacterial growth from selective media such as MacConkey, EMB, etc.

Note: (a) Before addition of reagents, observe GLU tube (positive or negative) for bubbles.

(b) The nitrate reduction and indole tests must be performed last since these reactions release gaseous products that in-

terfere with the interpretation of other tests on the strip. The plastic incubation lid should not be replaced after the addition of these reagents.

(b) Add the reagents to TDA and (VP) tubes. If positive, the TDA reactions will be immediate, whereas the (VP) reaction may be delayed up to 10 minutes.

(c) The Kovacs reagent should then be added to the IND tube.

(d) The Nitrate Reduction test should be performed on all oxidase-positive organisms (Figure 41-9). The reagents should be added to the GLU tube after the Kovacs reagent has been added to the IND tube.

(4) After all reactions have been recorded on the report sheet and satisfactory identification has been made, the entire incubation unit must be autoclaved, inciner-

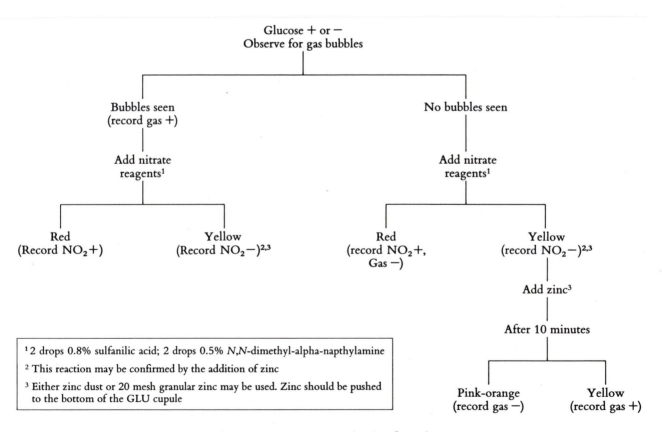

[1] 2 drops 0.8% sulfanilic acid; 2 drops 0.5% *N,N*-dimethyl-alpha-napthylamine

[2] This reaction may be confirmed by the addition of zinc

[3] Either zinc dust or 20 mesh granular zinc may be used. Zinc should be pushed to the bottom of the GLU cupule

Figure 41-9 Nitrate reduction flow chart.

ated, or immersed in a germicide prior to disposal.

6. IDENTIFICATION

On the report sheet you will find numerical values for each test. Add up all the numbers for each section (Figure 41-8F) and enter the number on the report sheet. Look up the 7-digit code generated in the API 20E Analytical Profile Index. Record the identity of the organism.

Identifying Unknown Bacterial Cultures: Miniaturized Multitest Methods for Identifying Enterics

Name _____

Desk No. _____

Record the reactions of your API strip or Enterotube II and the identity of the organism.

api 2OE® System

Reference Number _____ Patient _____ Date _____

Source/Site _____ Physician _____ Dept./Service _____

	ONPG 1	ADH 2	LDC 4	ODC 1	CIT 2	H₂S 4	URE 1	TDA 2	IND 4	VP 1	GEL 2	GLU 4	MAN 1	INO 2	SOR 4	RHA 1	SAC 2	MEL 4	AMY 1	ARA 2	OXI 4
5 h																					
24 h																					
48 h																					
Profile Number																					

	NO₂ 1	N₂ GAS 2	MOT 4	MAC 1	OF-O 2	OF-F 4
5 h						
24 h						
48 h						
Additional Digits						

Additional Information

Identification

Report when VP is included in diagnosis.

ENTEROTUBE® II*

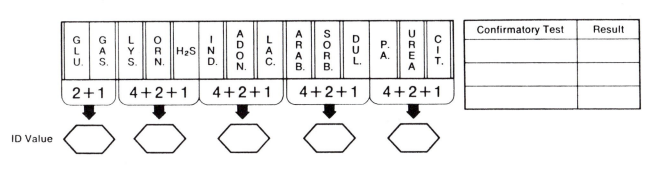

Confirmatory Test	Result

ID Value

Culture Number, Case Number or Patient Name _____ Date _____ Organism Identified _____

*VP utilized as confirmatory test only.

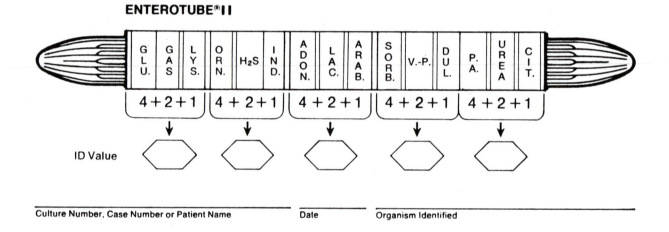

42. *Morphological Identification of Some Unusual Bacteria*

Even the smallest, simplest living things exhibit great complexity and variation. Not all bacteria are simple, rods, cocci, and spirals, nor is bacterial replication limited to binary fission. The diversity of the organisms studied in this exercise will be evident not only in morphology but also in size, motility, and mechanisms of multiplication. These microorganisms include forms that have stalked, sheathed, prosthecate, spiral, and filamentous morphologies. Their dimensions range from 0.5 to 200 μm. The life cycles of a few involve swarmers, and others form spores or cysts. They may be gram-positive or gram-negative or very difficult to stain.

Your textbook contains additional information illustrating habitats, complex interrelationships, and the ecological significance of these bacteria.

PROCEDURE

Several unidentified cultures will be provided for groups of students. Observe as much as possible about these organisms with the phase microscope (Exercise 4) and the gram stain (Exercise 13). Among the cultures provided may be *Ancalomicrobium, Caryophanon, Caulobacter, Myxococcus, Hypomicrobium, Rhodospirillum, Sphaerotilus, Streptomyces,* and *Anabaena*, a blue-green bacterium.

QUESTIONS

1. Which of the procaryotic organisms in this exercise is largest?
2. Which of the organisms in the exercise is smallest?
3. Which genera produce motile swarmers?
4. Which organisms are pigmented?
5. Which organisms produce spores?
6. Which organism produces a characteristic earthy smell?

Morphological Identification of Some Unusual Bacteria

Record the likely identity of the genera of unidentified cultures provided to you. Indicate the microscopic or other evidence that led you to your conclusions.

ISOLATION OF BACTERIA   In the following exercises you will seek and isolate members of specified groups of bacteria. For a successful isolation you must understand the habitats and the distinguishing characteristics of the bacterial group you seek. The best source of a bacterium is its usual habitat, for example, soil for bacilli and fermenting foods for the lactics. To select and enrich for a particular bacterial group, you must know the characteristics of its members that distinguish them from countless other microorganisms in the source material. Thus, to isolate a *Bacillus* species we take advantage of the heat resistance of this genus' endospores and select by heating our source material. Once isolated, the proof of the identity of a genus or species also requires in-depth knowledge of the various groups of bacteria. The attempt to find and isolate particular microbial groups is one of the more challenging and exciting problems of microbiology.

43. Bacillus *Isolation*

Members of this genus are gram-positive, aerobic or facultatively anaerobic, catalase-positive, endospore-forming rods (Figure 43-1 and Color Plate 2). Common sources are air, water, and soil.

As in most isolations, the nature of the source material and its environment are factors in determining the microbes most likely to be isolated. For example, the ambient temperature of the source plays a selective role in determining the optimum growth of your isolate. In *Bacillus* isolation, although soil is a source of either mesophiles or thermophiles, material from a hot manure pile is a better source of thermophilic strains.

For isolating *Bacillus*, the source material is heated to destroy vegetative cells. However, endospores not only survive, but many are stimulated to germinate by heating (termed **heat shocking**). Subsequent incubation of the heated sample enables the population to grow, and the resulting increase in cells of endospore formers facilitates the isolation. The temperature of incubation selects for the optimum growth temperature of the isolate. Thus, if you wish to isolate a thermophilic strain of *Bacillus*, use an incubation temperature of 55°C or higher.

Isolation of pure strains is achieved by streak plating, and the plates are incubated aerobically to inhibit growth of the anaerobic endospore-forming clostridia. Isolates are made from colonies composed of catalase-positive, motile, gram-positive

Figure 43-1 Photomicrograph of a *Bacillus* with an endospore and parasporal body. (Photo by J.J. Lee.)

rods and confirmed as *Bacillus* by the subsequent demonstration of endospore formation (see Figure 43-1).

PROCEDURE

Enrichment

1. Add the source material to a tube of nutrient broth and heat in a hot water bath at 80°C for 10 minutes (or steam at 100°C for 5 minutes).
2. Incubate the heated culture at 30°C, 35°C, or 55°C (depending on whether you wish to isolate a mesophile or thermophile) until the next laboratory period.

Isolation

3. After incubation streak a portion of the enrichment culture on a nutrient-agar plate and incubate at the temperature of your choice until the following laboratory period.
4. Examine several well-isolated colonies on the streak plate for their catalase reaction (Exercise 38); make a wet mount and examine microscopically their morphology and motility.
5. When you have located a colony of catalase-positive rods, inoculate a tube of nutrient broth. Emulsify a small bit of the colony in broth and streak a nutrient agar plate. Incubate the broth and plate cultures at the temperature of your choice for 24 hours.

Confirmation

6. After incubation examine the streak plate for the presence of an isolated colony having the colony characteristics of the original isolate. Also determine the catalase reaction of an isolated colony.
7. Prepare gram stains and endospore stains (Exercise 15) from the nutrient broth and isolated colonies on the agar plate. Examine microscopically. If the isolate is a catalase-positive, gram-positive rod, proceed with the characterization. Since all isolates do not routinely produce endospores on usual laboratory media, you may not have observed endospores at this step. Most *Bacillus* strains will form endospores within eighteen hours when grown on media containing added manganese.

Characterization

8. Inoculate the following media from your nutrient broth culture: a manganese-agar slant, a BCP glucose-agar shake, and starch- and casein-agar plates. Incubate at the desired temperature. Also inoculate four tubes of nutrient broth, incubating one tube each at 55°C, 45°C, 30°C, and 4°C.
9. After 24 hours incubation prepare an endospore stain from the manganese-agar culture. Examine microscopically for endospore formation. If present, note their size, shape, and cellular location.
10. After 48 hours incubation:
    (a) Observe the BCP glucose shake for glucose fermentation and the oxygen relationship of your isolate.
    (b) Test the starch- and casein-agar plates for starch (Exercise 35) and casein (Exercise 36) hydrolysis.
    (c) Observe the nutrient broth tubes for growth at the various temperatures to determine the optimum and limiting temperatures of your isolate.
11. Based on these observed characteristics can you identify the species of *Bacillus* isolated?

QUESTIONS

1. Why is the *Bacillus* enrichment medium heated?
2. Why is the *Bacillus* enrichment medium incubated before inoculating an agar plate for isolation?
3. Why is the *Bacillus* culture inoculated on a manganese slant?
4. When may a *Bacillus* culture seem to be catalase-negative?
5. What is a parasporal body?

44. *Pseudomonad Isolation*

Members of this group are gram-negative rods that are generally motile by polar flagella. The pseudomonads are strict aerobes though some species can grow anaerobically, using nitrate in lieu of O_2 as a terminal electron acceptor. They do not ferment but dissimilate carbohydrates and many other organic compounds oxidatively. Associated with their aerobic nature is the ability to oxidize aromatic amines, such as *p*-aminodimethylaniline, to form colored end products. The oxidation of aromatic amines is related to the presence of cytochrome *c* in their respiratory chain and is the basis of the so-called oxidase test, which is commonly used to identify the oxidase-positive pseudomonads. These bacteria are metabolically versatile and of widespread occurrence with sources including soil, pond water, sewage, eggs and egg shells, curing brine, raw milk, fish, shellfish, leafy vegetables, and poultry.

For this isolation you will take advantage of the resistance of pseudomonads to many antibiotics by plating the source material on agar medium that contains the antibiotics novobiocin, penicillin, and cycloheximide. Because penicillin and novobiocin inhibit gram-positive bacteria and cycloheximide suppresses fungi, and because incubation is aerobic, this procedure is highly selective for the pseudomonads. Pseudomonad colonies will be those that are composed of gram-negative, oxidase-positive rods that do not ferment glucose. You can further identify your pseudomonad isolate by its growth temperatures, pigment production, and phospholipase and gelatinase activities.

PROCEDURE
Enrichment

1. To increase the limited numbers of organisms in the source material, it is preincubated at 20°C for 24 hours. Solid sources such as eggs, fish, and poultry should be placed in phosphate-buffered dilution blanks, shaken, and incubated at 20°C for 24 hours. This serves to wash the bacteria from the solid materials and facilitates their distribution during streak plating.

2. After preincubating, streak the source material onto a novobiocin–penicillin–cycloheximide agar plate. Incubate at 20°C (or other temperature stated by your instructor) for 24–48 hours.

Isolation

3. Make a wet mount of one or more colonies and examine microscopically. When you have located a colony of motile rods, perform an oxidase test on a portion of the colony using the filter paper method (Exercise 38). An oxidase-positive colony, along with the proper cell morphology, is presumptive evidence for a pseudomonad.

4. When you have located a colony of oxidase-positive rods, inoculate a tube of nutrient broth from it. Also emulsify a small bit of the colony in sterile broth and streak a nutrient-agar plate. Incubate both at 20°C for 24 hours.

Confirmation

5. Examine the nutrient-agar confirmation plate for the presence of a pure culture and typical colony characteristics.

6. From the nutrient broth culture prepare and examine a gram stain to confirm that the positive-oxidase reaction of Step 3 is produced by a gram-negative rod and not an oxidase-positive strain of the gram-positive genus *Bacillus*.

7. From the nutrient broth culture inoculate two tubes of oxidation–fermentation (OF) agar by stabbing the agar nearly to the bottom of the tube. Cover one tube with vaspar and incubate both tubes at 20°C for 48 hours or longer (Exercise 35).

8. Observe the OF tubes for acid production from glucose as indicated by the bromthymol blue indicator (yellow). The oxidative pseudomonads grow and produce slight amounts of acid only at the surface of the uncapped tube but not in the capped or fer-

mentation tube test. Gas production is manifested by bubbles trapped in the agar, and motility is demonstrated if turbidity extends beyond the stab line.

You can further characterize your pseudomonad (see Table 44-1) by determining if growth occurs at 5°, 35°, and 41°C and by testing for nitrate reduction, phospholipase C production on lipid hydrolysis agar (Exercise 37), gelatin hydrolysis (Exercise 36), and the production of pyocyanin and fluorescein pigments. Elaboration of these pigments is a distinctive characteristic of certain strains of the genus *Pseudomonas*.

You can demonstrate the production of these pigments by using special agars including Tech (*Pseudomonas* P agar) and Flo (*Pseudomonas* F agar).* Pyocyanin is a blue-green water-soluble pigment. On Tech agar this pigment appears within 6 or 7 days. Fluorescein production is enhanced by culturing on Flo agar because of this medium's limiting level of ferric iron. Fluorescein-producing cultures on Flo-agar plates fluoresce when the plate is opened and viewed under ultraviolet radiation. To avoid mistakes in interpreting pigment production, it is best to compare your isolate with positive and negative controls provided by your instructor.

---

*These agars are commercially available from Baltimore Biological Labs at Cockeysville, MD 21030, and Difco Labs, Detroit, MI 48232.

Table 44-1   Reactions of Some Common *Pseudomonas* species

Test	P. aeruginosa	P. fluorescens	P. putida
Growth at 4°C	−	+	±
Growth at 41°C	+	−	−
Pyocyanin production	+	−	−
Fluorescein production	+	+	+
Phospholipase production	−	+	−
Gelatin hydrolysis	+	+	−

QUESTIONS

1. Explain the rationale for using novobiocin–penicillin–cycloheximide agar for pseudomonad isolation.
2. Some oxidase-positive nonpseudomonads, for example, *Neisseria* and *Streptomyces*, may grow on the antibiotic-selective medium. How could you eliminate these in the isolation procedure?
3. Why is the fluorescein pigment produced in the Flo medium and not on plate count agar?
4. What enzyme does the oxidase test demonstrate?

Pseudomonad Isolation

Name _____

Desk No. _____

Describe the approach you used to isolate a pseudomonad. Include such information as source, rationale, and key tests used for identification. Use the Descriptive Chart on the back of this page to record reactions.

Pseudomonad Isolation

PART A: MORPHOLOGY

VEGETATIVE CELLS

Form	spheres, short rods, long rods, filaments, commas, spirals
Arrangement	singles, pairs, chains, fours, cubical packets, clusters
Capsules	absent, present

Staining reaction	gram- _____
Motility	present, absent

PART B: CULTURAL CHARACTERISTICS

COLONY

Medium: _____ Age: _____

Growth	slow, rapid, moderate
Form	punctiform, circular, rhizoid, irregular
Surface	smooth, rough, dry, moist, dull, glistening
Elevation	flat, raised
Edge	smooth, wavy, filamentous

SLANT

Medium: _____

Growth	scanty, moderate, abundant
Form	threadlike, beaded, spreading, rootlike
Elevation	flat, raised
Optical characters	opaque, translucent, iridescent
Color	_____ water-soluble, water-insoluble
Consistency	butyrous, viscid, brittle

BROTH

Medium: _____

Surface growth	ring, pellicle, none
Clouding	slight, heavy, none
Amount of sediment	abundant, scanty, none
Type of sediment	flaky, granular, viscid on agitation

TEMPERATURE OPTIMUM: _____ °C

PART C: PHYSIOLOGY

GELATIN STAB

Liquefaction	none, slow, moderate, complete

FERMENTATIONS

Glucose	acid, gas, negative
	Incubation time: _____
Lactose	acid, gas, negative
	Incubation time: _____
Sucrose	acid, gas, negative
	Incubation time: _____

PART D: ADDITIONAL DATA

SPECIFIC TESTS *(indicate positive or negative)*

Nitrate reduction
Oxidase
Lipid hydrolysis

Pigment production	yes _____ no _____
Color	_____
Fluorescence	yes _____ no _____

45. Staphylococcus *Isolation*

The family Micrococcaceae includes the genera *Staphylococcus and Micrococcus*. They are salt-tolerant, gram-positive, catalase-positive, nonmotile cocci. *Micrococcus* is a strictly aerobic genus with cells arranged in clusters, tetrads, or cubical packets. The genus *Staphylococcus* is facultatively anaerobic, produces acid and no gas from glucose, and is characteristically arranged in irregular clusters (Figure 45-1). Common sources of *Micrococcus* are soil, air, fresh water, and skin. *Staphylococcus* is found on skin and mucous membranes of warm-blooded animals and in some foods and curing brines.

A simple enrichment procedure for *Staphylococcus* is to culture the source material anaerobically in a medium containing 7% NaCl. The high osmotic concentration of this medium inhibits osmotically sensitive organisms. Anaerobic culture discourages the growth of aerobic *Bacillus* and *Micrococcus*. The culture is purified by streak plating from the enrichment broth and selecting a colony composed of organisms that are catalase-positive and gram-positive cocci and that ferment glucose anaerobically.

PROCEDURE

Enrichment

1. Inoculate the source material into 7% NaCl broth and seal the tube with vaspar.
2. Incubate at 30°C for 2 days.
3. After incubation, streak a portion of the enrichment culture on a plate-count agar plate and incubate at 30°C for 2 days.

Isolation

4. Examine several isolated colonies on the streak plate for their catalase reaction (Exercise 38) and their morphology as seen microscopically in a wet mount.
5. When you have located a colony of catalase-positive cocci, inoculate a tube of nutrient broth. Emulsify a small bit of the colony in sterile broth and streak a plate-count agar plate. Incubate both at 30°C for 24 hours.

Confirmation

6. After incubation, examine the streak plate for purity to assure that a single colony type was transferred. Prepare and examine a gram stain from the broth culture.
7. If your isolate is composed of gram-positive cocci in clusters, inoculate a bromcresol purple glucose shake from the broth culture and incubate at 30°C for 5 days.
8. Observe the shake culture for glucose fermentation. If your isolate of catalase-positive, gram-positive cocci ferments glucose forming only acid is a facultative anaerobe, you have isolated a species of *Staphylococcus*. You can further identify the species (see Table 45-1) by testing for mannitol fermentation (Exercise 35), novobiocin sensitivity (Exercise 71), and deoxyribonuclease and co-

Figure 45-1   Scanning electron micrograph of *Staphylococcus* (10,000 ×). (Photo by J.J. Lee.)

Table 45-1   Reactions of the Staphylococci

Test	S. aureus	S. epidermidis	S. saprophyticus
Coagulase activity	+	−	−
DNase activity	+	−	−
Novobiocin sensitivity[a]	S	S	R
Mannitol fermentation	+	−	−

[a]S = sensitive; R = resistant

agulase activities (see also Exercise 70C for API Staph Ident tests). Production of DNase, coagulase, acid from mannitol, and yellow or orange pigment are correlated with pathogenic *Staphylococcus aureus*.

QUESTIONS

1. What other common genera can grow in the *Staphylococcus* enrichment medium?
2. How many hours does a clinical microbiologist allow for the appearance of a positive coagulase test?
3. What is the substrate for a coagulase reaction?
4. How should the bromcresol purple glucose shake inoculated with *Staphylococcus* appear after 5 days?

Staphylococcus Isolation

Describe the approach you used to isolate a *Staphylococcus*. Include such information as source, rationale, and key tests used for identification. Use the Descriptive Chart on the reverse side of this page to record reactions.

DESCRIPTIVE CHART

Staphylococcus Isolation

PART A: MORPHOLOGY

VEGETATIVE CELLS

Form	spheres, short rods, long rods, filaments, commas, spirals
Arrangement	singles, pairs, chains, fours, cubical packets, clusters
Capsules	absent, present

Staining reaction   gram-_____

PART B: CULTURAL CHARACTERISTICS

COLONY

Medium: _____ Age: _____

Growth	slow, rapid, moderate
Form	punctiform, circular, rhizoid, irregular
Surface	smooth, rough, dry, moist, dull, glistening
Elevation	flat, raised
Edge	smooth, wavy, filamentous

SLANT

Medium: _____

Growth	scanty, moderate, abundant
Form	threadlike, beaded, spreading, rootlike
Elevation	flat, raised
Optical characters	opaque, translucent, iridescent
Color	_____ water-soluble, water-insoluble
Consistency	butyrous, viscid, brittle

BROTH

Medium: _____

Surface growth	ring, pellicle, none
Clouding	slight, heavy, none
Amount of sediment	abundant, scanty, none
Type of sediment	flaky, granular, viscid on agitation

OXYGEN RELATED

_____: in _____ (medium)

TEMPERATURE OPTIMUM:_____°C

PART C: PHYSIOLOGY

GELATIN STAB

Liquefaction   none, slow, moderate, complete

FERMENTATIONS

Glucose	acid, gas, negative

Incubation time: _____

Lactose	acid, gas, negative

Incubation time: _____

Sucrose	acid, gas, negative

Incubation time: _____

Mannitol	acid, gas, negative

Incubation time: _____

PART D: ADDITIONAL DATA

GROWTH ON BLOOD AGAR
   Hemolysis   yes _____ no _____

SPECIFIC TESTS *(indicate positive or negative)*

Catalase
Citrate utilization
Coagulase
Deoxyribonuclease
Novobiocin sensitivity

46. *Lactic Acid Bacteria Isolation*

The lactic acid bacteria are gram-positive cocci or rods that ferment carbohydrates primarily to lactic acid. They are catalase-negative, microaerophilic or facultatively anaerobic bacteria that are common in dairy products, fermented foods, and the mouth and gastrointestinal tracts of humans and animals.

For isolation, one can use the selective advantage that these bacteria lack cytochromes and are therefore insensitive to inhibitors such as azide. Microbes whose growth depends on cytochrome-linked respiration are inhibited. By incorporating this inhibitor into a nutritional medium that contains a fermentable carbohydrate as an energy source, one enriches for the growth of lactics. Colonies of lactic acid bacteria growing on this medium can be identified as catalase-negative, gram-positive, non-endospore-forming cocci or rods.

PROCEDURE

Enrichment

1. Inoculate two tubes of steamed azide yeast-extract broth with material from two different sources. (If you want to isolate lactobacilli, inoculate tubes of all-purpose APT-azide broth.) It is to your advantage to inoculate with a relatively large amount of source material; for example, if you use raw milk, use 5 loopfuls.
2. Layer both broth cultures with vaspar and incubate both tubes for 24 or 48 hours at a temperature that approximates the environmental temperature of the source material.

Isolation

3. Examine the enrichment culture microscopically using a wet mount or a gram stain (Exercise 13). Using the culture with the greatest growth, streak an APT-calcium carbonate agar plate. The calcium carbonate in this agar acts as a buffer and also permits selection of acid-producing colonies. The acid causes a zone of clearing around these colonies due to the solubilization of the cal-

cium carbonate. Incubate the streak plate for 48 hours at the same temperature used for the enrichment.
4. After incubation, perform catalase tests (Exercise 38) on cells from isolated colonies until a catalase-negative colony is located. From this colony inoculate a tube of strep-base or APT broth. Emulsify a small bit of the colony in sterile broth and streak an APT-agar plate. Incubate both at the appropriate temperature.

Confirmation

5. After 24 hours of incubation prepare a gram stain from the broth culture. Examine microscopically to determine the cell morphology and arrangement and the gram reaction of your isolate (Figure 46-1).

Figure 46-1 Scanning electron micrograph of chains of *Streptococcus pyogenes* (10,000 ×). (Photo by J.J. Lee.)

6. After 48 hours' incubation of the APT streak plate, check for culture purity and colony characteristics. Also recheck the catalase reaction of one or more isolated colonies.

7. From the broth culture inoculate a BCP–APT agar shake. Layer the agar shake with vaspar. If your isolate is a rod, also inoculate a manganese-agar slant. Incubate both tubes at your isolation temperature for 48 hours.

Table 46-1 Differentiation of the Lactic Acid Bacteria

Genus	Cell Form and Arrangement	Fermentation
Streptococcus	Cocci in chains	Homo
Leuconostoc	Cocci in chains	Hetero
Pediococcus	Cocci in tetrads	Homo
Lactobacillus	Rods	Homo or hetero

8. Observe the APT-agar shake for gas formation. Heterofermentative organisms will accumulate gas under the vaspar cap.

9. If your isolate is a rod, prepare and observe a spore stain from the manganese slant to assure that the isolate is not a spore former.

10. Based on its morphological and fermentative characteristics, identify the genus of lactic bacteria you have isolated (Table 46-1).

QUESTIONS

1. Why are media for lactic isolations steamed before use?
2. What is the purpose in using a vaspar cap with the lactic enrichment?
3. When is the capillary technique for catalase most appropriate?
4. Why is calcium carbonate included in some lactic-plating media?

Lactic Acid Bacteria Isolation

Name _____

Desk No. _____

Describe the approach you used to isolate a lactic acid bacterium. Include such information as a source, rationale, and key tests used for identification. Use the Descriptive Chart on the reverse side of this page to record reactions.

DESCRIPTIVE CHART

Lactic Acid Bacteria Isolation

PART A: MORPHOLOGY

VEGETATIVE CELLS

Form   spheres, short rods, long rods, filaments, commas, spirals

Arrangement   singles, pairs, chains, fours, cubical packets, clusters

Capsules   absent, present

Staining reaction   gram- _____

Motility   present, absent

Spore formation on
Manganese slant   present, absent

PART B: CULTURAL CHARACTERISTICS

COLONY

Medium: _____ Age: _____

Growth   slow, rapid, moderate

Form   punctiform, circular, rhizoid, irregular

Surface   smooth, rough, dry, moist, dull, glistening

Elevation   flat, raised

Edge   smooth, wavy, filamentous

SLANT

Medium: _____

Growth   scanty, moderate, abundant

Form   threadlike, beaded, spreading, rootlike

Elevation   flat, raised

Optical characters   opaque, translucent, iridescent

Color   _____ water-soluble, water-insoluble

Consistency   butyrous, viscid, brittle

BROTH

Medium: _____

Surface growth   rin, pellicle, none

Clouding   slight, heavy, none

Amount of sediment   abundant, scanty, none

Type of sediment   flaky, granular, viscid on agitation

Growth in APT agar shake

TEMPERATURE OPTIMUM: _____

PART C: PHYSIOLOGY

GELATIN STAB

Liquefaction   none, slow, moderate, complete

LITMUS MILK

Reduction   slight, moderate, complete, none

Incubation time: _____

FERMENTATIONS

Glucose   acid, gas, negative

Incubation time: _____

Lactose   acid, gas, negative

Incubation time: _____

Sucrose   acid, gas, negative

Incubation time: _____

PART D: ADDITIONAL DATA

APT-CaCO$_3$ PLATE

Growth   present, absent

Cleaving   yes _____ no _____

SPECIFIC TESTS *(indicate positive or negative)*

Catalase

Bacterial Variation, Mutation,
and Recombination

Observation of pure bacterial cultures might lead you to believe that all cells of a culture are alike, but a cell population contains mutants whose characteristics differ from those of the parent cells. During the growth of a bacterial culture, mutations occur at very low frequencies, perhaps in one cell out of 10 million. Mutant cells differ from the parent culture in some way, such as failure to ferment a certain sugar, loss of the ability to form a toxin, or loss of resistance to a certain drug. The altered characteristic is stable and appears in all cells that develop from the mutant. Permanent, inherited variation of this sort reflects a change in the genetic apparatus, the **genotype**, of the microbe. The *observed* difference in morphological and physiological characteristics resulting from alteration of genetic material is called **phenotypic variation**. Phenotypic variation is usually the expression of a change in the genotype. However, observed variation in the cell's activities may be due solely to influences in the environment in which a microbial population is grown. For example, the cell may produce additional enzymes that synthesize needed nutrients such as amino acids that are usually in short supply. If these nutrients become readily available in the cell's environment, however, one type of regulation (feedback inhibition) prevents their continued synthesis. In contrast to a mutation that appears in a single cell and its progeny, feedback inhibition and other regulatory mechanisms occur in all cells of a culture in response to environmental changes.

Changes in the genetic makeup of bacteria can also result from the transfer of genetic material from a donor to a recipient cell. Three types of genetic transfer in bacteria are transformation, conjugation, and transduction. In **transformation**, free or "naked" DNA from a lysed donor cell is incorporated into a recipient cell. **Conjugation** is the transfer of DNA with close physical contact between the donor and recipient cells. **Transduction** is the transfer of genetic material to a bacterium by a bacterial virus. Transfer of genetic material by transformation, conjugation, or transduction can be followed by a recombination of the transferred DNA with the recipient's genome; but only a part of the donor's genome is transferred to the recipient cell.

These forms of genetic exchange appear to occur at a low frequency and have been observed among only a relatively small number of bacterial species. Even so, genetic transfer and recombination can change characteristics of bacterial cells and their progeny which affect biochemical activities, ability to cause disease, and resistance to antibiotics.

47. *Bacterial Variation*

The microbial cell can be regarded as a great complex of interlacing chemical activities that exhibit considerable flexibility in response to the environment. Thus, capsules of *Alcaligenes viscolactis* are produced at low temperatures but not at high temperatures, though the organism grows at the higher temperatures. Variation in nutritional requirements occurs among microbial cells. For example, both the colonies and cells of the "minute" streptococci increase in size when a small amount of CO_2 is supplied to the culture. *Lactobacillus casei* can be grown at pH 5 without adding the B-vitamin pyridoxine to the medium, but at pH 7 pyridoxine must be added. Most of the reactions of the variant forms are completely unknown, but it is clear that they pertain to only a part of the cells' complex activities.

This exercise illustrates temporary phenotypic variation in the pigmentation of *Serratia marcescens* when growing at different temperatures.

PROCEDURE

1. Prepare two petri plates of nutrient agar. Using the techniques outlined in Exercise 9, prepare two streak plates of *Serratia marcescens*.
2. Incubate one plate at 25°C or room temperature and the other at 37°C until the next laboratory period.
3. Observe the effect on the temperature of incubation on pigment formation. Record the results on your report sheet.

QUESTIONS

1. How would you explain the results of this experiment?
2. Does phenotypic variation without genotypic variation occur in higher forms of life?

Bacterial Variation

Name _____

Desk No. _____

Record the results of pigment production by *Serratia marcescens*.

Incubation temperature	Color
25°C	
37°C	

48. *Enzyme Induction*

Regulation of enzyme synthesis is another aspect of bacterial adaptation to environment. Constitutive enzymes such as those necessary for the metabolism of glucose are produced under all conditions of growth, but many enzymes are inducible; that is, they are produced only in the presence of a particular substrate or in particular environmental conditions. For example, alkaline phosphatase is a repressed enzyme, which is produced only when the cell is starved for inorganic phosphate, and it is functional only in alkaline conditions.

An essential inorganic nutrient such as phosphate can be growth limiting. In such a case, the cells respond physiologically to the concentration of inorganic phosphate in the medium by making or repressing production of alkaline phosphatase, an enzyme that hydrolyzes organic phosphate esters to liberate the inorganic phosphate.

Table 48-1   Phosphate Content of Media for Measurement of Alkaline Phosphatase

	Amount of Reagent (ml)		
Tube Number	100 µg/ml Na$_2$HPO$_4$	Phosphate-free TRIS Buffer	Final Phosphate Concentration (µg/ml)
0	0	5	0
1	0.05	5	1
2	0.075	4.9	1.5
3	0.1	4.9	2
4	0.15	4.85	3
5	0.2	4.8	4
6	0.3	4.7	6
7	0.4	4.6	8
8	0.6	4.4	12

PROCEDURE

First Laboratory Period

1. Each pair of students will be given a bottle of double strength minimal medium. This medium contains all the inorganic salts required (NH$_4^+$, SO$_4^{2-}$, and so on) for growth of *Escherichia coli*, but no phosphate. Glucose is present as a carbon and energy source. Aseptically add 5 ml of 2✕ (double strength) medium to each of 9 sterile screw-capped test tubes. A sterile solution containing 100 µg/ml phosphate will be provided.

2. Prepare a series of sterile tubes as illustrated in Table 48-1.

3. Using one drop of culture grown in a medium that contains 4 µg/ml phosphate, inoculate the series of tubes. This should insure that your cultures will have stopped growing shortly before the start of the next laboratory period (about 48 hours).

PROCEDURE

Second Laboratory Period

Remove the caps from the tubes (this procedure is not aseptic).

1. Mix the contents well and wipe the outsides of the tubes.

2. Determine the levels of growth in the cultures by measuring their absorbances in a spectrophotometer at 650 nm. Use a tube of water to zero the instrument.

Assay of Alkaline Phosphatase

Alkaline phosphatase activity can be measured using phosphate esters. A convenient substrate is *p*-nitrophenyl phosphate (pNPP). This substance is colorless, but the *p*-nitrophenol that is liberated when the phosphate group is removed by hydrolysis is bright yellow. Therefore, by using a spectrophotometer and the blue filter (425 nm), you can ascertain the activity of an enzyme preparation by measuring the yellow color formed after a standard time of incubation.

Proceed in the assay of alkaline phosphatase as follows:

1. From a dropping bottle, add one drop of toluene to each culture. *Caution*: Toluene is flammable. Incubate 10–15 minutes in a 37°C water bath; this destroys the perme-

ability barrier of the whole cell so that enzyme and substrate can react freely.

2. Prewarm the substrate solution (20 mg/ml pNPP) to 37°C in a water bath. The reaction should be carried out at a constant temperature because enzyme activity is highly temperature dependent.

3. Exact timing is required in this step, so you should work at the water bath. Add 0.1 ml of substrate solution to each tube (Table 48-1) in order, 0 through 8, at equal time intervals. Exactly 5 minutes after adding substrate to a tube, stop the enzyme reaction by adding 0.2 ml of 1 M K_2HPO_4 to them with a 5-ml pipet. This final concentration of phosphate is strongly inhibitory to alkaline phosphatase production.

4. Centrifuge the tubes to remove cells from suspension. Be careful not to disturb the pel-

let. Measure absorbance of the supernatant in a spectrophotometer at 425 nm.

5. Using the charts on the report sheet,
(a) draw a graph relating final cell yield to the concentration of inorganic phosphate in the medium; and
(b) plot and draw a graph of the alkaline phosphatase activity divided by the growth (A) versus the phosphate concentration.

QUESTIONS

1. Does *E. coli* make more alkaline phosphatase per cell when the medium contains little or excess inorganic phosphate? Why might this prove helpful to the organism?

2. Is the relationship between final yield and inorganic phosphate linear? Should it be? Explain.

Name _____

Desk No. _____

Plot the OD at 42.5 nm versus the inorganic phosphate concentration.

OD (A at 42.5 nm)

1.0

0.5

0   4   8   12

Phosphate Concentration
(μg/ml)

Plot the alkaline phosphatase activity divided by the growth absorbency versus the phosphate concentration.

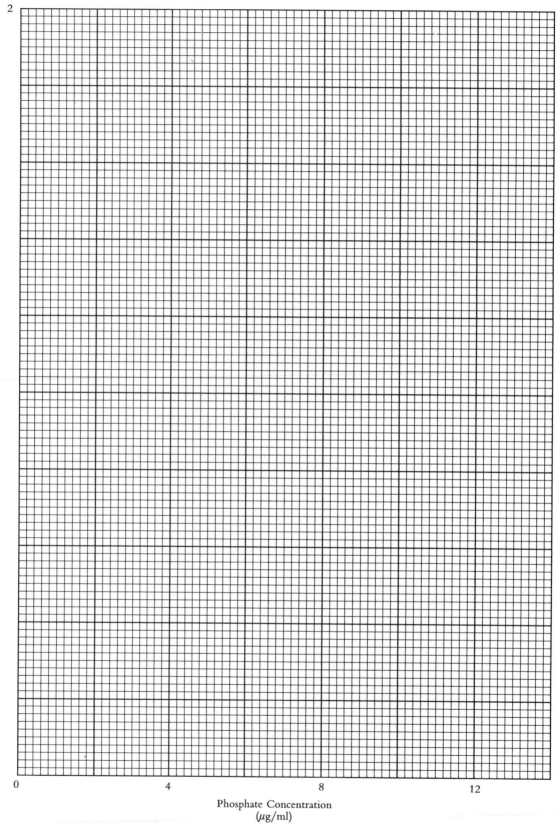

49. *Genotypic Change*

BACTERIAL MUTATION: ISOLATION OF A STREPTOMYCIN-RESISTANT MUTANT

Because of the low frequency of mutation and because special methods must be used to select mutants, demonstrating mutant cells is not an easy task. However, some manipulations of the physical or chemical environment can increase the frequency of mutants, for example, radiation with x rays or ultraviolet light (see Exercise 31). If a mutant cell, by virtue of the change that it has undergone, is somehow especially suited to the environment in which it is formed, its growth can outstrip that of its parent culture, and it may become dominant.

Spontaneous mutants that are resistant to an antibiotic are easily detected because they grow in concentrations of the antibiotics that would inhibit the growth of normal organisms. This exercise employs the gradient-plate method to isolate and select streptomycin-resistant mutants.

PROCEDURE

1. Prepare a gradient-agar plate (Figure 49-1A).
   (a) Tilt a sterile petri dish by placing a glass rod or wooden stick, approximately 1.6 mm in diameter, under one side. Pour a tube of melted trypticase-soy agar into the petri dish and allow the agar to harden, slanting the plate just enough to cover the entire bottom surface.
   (b) After the trypticase-soy agar has solidified, remove the prop and set the plate flat on the table. To a second tube of melted and tempered trypticase-soy agar, add 0.1 ml of streptomycin. Mix the tube contents well and pour the melted agar onto the surface of the gradient-agar plate (Figure 49-1B).

2. Pipet approximately 0.3 ml of *Staphylococcus aureus* culture onto the agar surface. *Caution*: This is a pathogen. Using a sterile glass spreader (sterilized by dipping in 95% alcohol, flaming, and cooling on the sterile surface of the gradient plate), spread the culture over the agar surface (see Figure 17-2). Incubate at 37°C until the next laboratory period.

3. Observe for the development of resistant colonies in the areas of higher streptomycin concentration. With a sterile loop make a suspension of cells in broth from colonies growing in the region of highest antibiotic concentration. From this broth suspension, inoculate a loopful of tubes of trypticase-soy broth containing 0.01 and 0.05 milligrams of streptomycin per tube.

A   B

1/16″ (1.6 mm)   Nutrient agar   Streptomycin agar   Nutrient agar

Figure 49-1   Preparation of a gradient-agar plate *A*: Place a 1.6 mm block under a petri plate and then add nutrient agar. *B*: After the plate cools, remove the block and then add the streptomycin agar.

Using the original strain of *S. aureus*, your instructor will inoculate another set of tubes to serve as controls. Incubate all tubes at 37°C for 48 hours.

4. Observe the tubes for growth and compare with the *S. aureus* control tubes. Record the results on the report sheet.

QUESTIONS

1. What are streptomycin-dependent mutants? How would you modify this exercise to obtain such mutants?
2. What methods could be used to increase the frequency of appearance of the mutant cells?

REPORT 49

Genotypic Change

Name _____

Desk No. _____

Record the level of growth, using a scale from 0 (no growth) to ++++ (heavy growth).

Inoculum	Streptomycin		
	0 mg	0.01 mg	0.05 mg
Original *Staphylococcus aureus*			
Resistant *Staphylococcus aureus*			

50. *Plasmid-Mediated Lactose Fermentation by* Streptococcus lactis

Many of the characteristics of microbes, for example, some forms of antibiotic resistance, are coded for by elements of extrachromosomal DNA termed **plasmids**. A high percentage of genetic transfer among bacteria also involves plasmids that can be passed from one cell to another under certain conditions.

In 1937, J. Sherman found that cultures of many lactic acid bacteria lost the ability to use lactose after prolonged cultivation in growth media that lacked it. Because it was generally accepted at that time that bacteria did not have genes, there was no explanation for this variation. With the advent of bacterial genetics and molecular biology, it was found that the genes for lactose-catabolizing enzymes in lactic acid bacteria were often on plasmids, which can be easily lost by the bacteria. When a bacterial cell loses a plasmid, the characteristics coded for by the plasmid genes are also lost.

We will demonstrate this phenomenon of loss of ability in a *Streptococcus lactis* culture, abetting the variation by using two agents that are known to "cure" plasmids: elevated temperature and acriflavin. Elevated temperature (in this case $37-40°C$) affects plasmid replication, apparently by interfering with their attachment to the cell membrane, which is necessary for replication. Acriflavin is a planar molecule that stacks between the DNA bases (called *intercalation*); it affects plasmid replication more than chromosomal replication. Acriflavin is also a strong frame-shift mutagen, so we must consider the possibility that it has caused a mutation rather than plasmid loss when it causes loss of a given property. We examine the ability of these two agents, together and separately, to cause the loss of lactose-catabolizing ability. We will use **replica plating** on pH indicator media for the detection of loss of the ability to catabolize lactose.

PROCEDURE

First Laboratory Period

1. You will be supplied with fresh cultures of *Streptococcus lactis* (ATCC 7962, or 11454), which have the ability to use lactose.

Transfer the culture into tubes of a rich, well-buffered medium called M-17 broth, which includes glucose as the energy source. You will use four different treatments. Label and treat four tubes of M-17 as follows:
(a) Control (No additions; incubate at $30°C$)
(b) Heat (No additions; incubate at $40°C$)
(c) Acriflavin (Add acriflavin to a final concentration of $1\ \mu g/ml$; incubate at $30°C$)
(d) Heat + acriflavin (Add acriflavin to a final concentration of $1\ \mu g/ml$; incubate at $37°C$)
2. Inoculate each tube with a loopful of the *S. lactis* culture (Figure 50-1).
3. *Caution*: Put on disposable gloves. *Acriflavin is a strong mutagen and suspected carcinogen. Use extreme care when handling.* Do not mouth pipet!! Dispose of contaminated materials as directed by your instructor. Carefully pipet 0.1 ml of a $100\ \mu g/ml$ solution of acriflavin (filter sterilized) into tubes (c) and (d).
4. Incubate the cultures at their respective temperatures ($30°$, $37°$, and $40°C$).
5. Examine the culture after 24 and 48 hours. When they are fully grown (10^9 cells/ml), place the cultures into the refrigerator until the next period. Your course assistant or instructor may do this.

Second Laboratory Period

1. You will need to test individual colonies in the next step, so you must serially dilute the broth cultures before you plate them. Make the serial dilutions by aseptically pipetting 0.1 ml of each of your labeled cultures (a–d) into a fresh tube of 10 ml of medium (10^{-2} dilution).
2. Thoroughly mix the 1×10^{-2} dilution in a vortex mixer.
3. Take 0.1 ml of the 10^{-2} dilutions and aseptically pipet it into 10 ml of fresh medium (1×10^{-4} dilution). Vortex vigorously.
4. Repeat the procedures so that you obtain

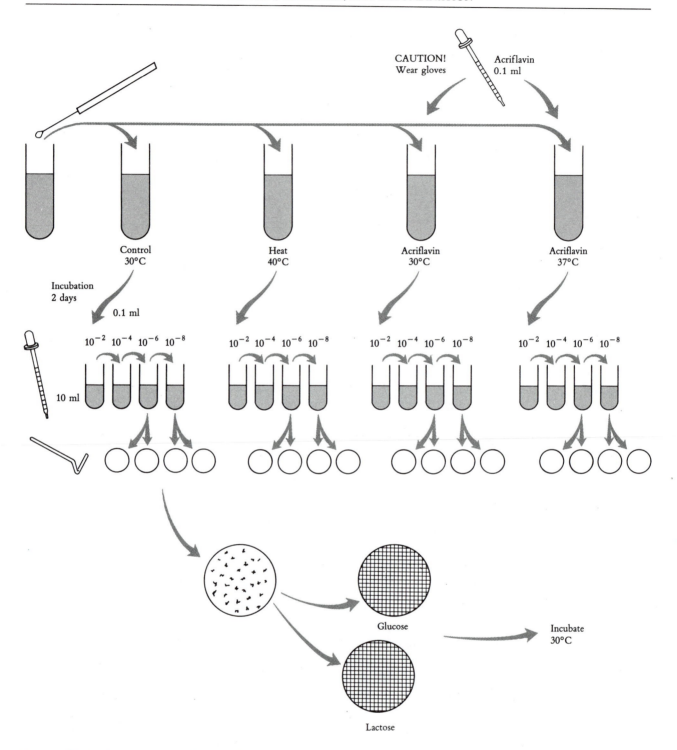

Figure 50-1   Flow chart of the procedure used to demonstrate the curing of plasmids by elevated temperature and acriflavin in *Streptococcus lactis*.

10^{-6} and 10^{-8} dilutions from each of the original four treatment cultures.

5. For each of the four treatments, prepare four M-17 petri spread plates as follows: Pipet 0.1 ml from the 10^{-6} dilution onto two of the M-17 agar plates, and pipet 0.1 ml from the 10^{-8} dilution onto two other M-17 agar plates. Spread with an alcohol sterilized glass rod.

6. Incubate at 30°C.

Third Laboratory Period

Five days after the second laboratory period, you will patch colonies that have grown on M-17 agar using toothpicks. You will transfer individual colonies (clones) onto BCP (bromcresol purple) agar containing either glucose or lactose as the fermentable sugar.

1. For each of the three treatments (b, c, and d), divide two glucose BCP and two lactose BCP plates with a marking pen into two sectors of roughly equal size and label them. Mark which plates are glucose, which are lactose so you can tell them apart, and which treatment each plate represents.
2. For each treatment, use sterile toothpicks to pick well-isolated colonies on the M-17 agar, and patch each colony pick onto both a lactose BCP plate and a glucose BCP plate (in that order) using a different toothpick until 50 colonies have been replica-plated. You should just touch the agar of the BCP plates lightly in the center of the sector.
3. Incubate the plates at 30°C for 18–24 hours. After that time, place the plates in the refrigerator until they can be scored.

Fourth Laboratory Period

1. Two days after the third laboratory period, score the plates for how many colonies from each treatment fermented lactose versus how many fermented glucose. Such colonies turn the agar yellow around them due to acid production from the carbohydrate.
2. Replate any colonies for which the results are ambiguous. Also, make wet mounts of any suspicious colonies (i.e., possible contaminants) and ascertain whether their morphology corresponds to that of *S. lactis*.

QUESTIONS

1. What are the rationales for the various steps in the experiment?
2. How can you prove that the loss of the ability to ferment lactose is not due to mutation?
3. What could be the economic importance of the loss of the lactose plasmid from *Streptococcus lactis*?

51. *Bacterial Conjugation*

Conjugation involves the transfer of DNA through cell-to-cell contact between a recipient (F⁻) and a donor (F⁺ or Hfr). Donor cells have a conjugative plasmid that codes for the formation of a surface structure, the sex pilus, which is involved in bridging the donor and a recipient and the transfer of DNA.

An Hfr male strain donates its genes to the recipient in a linear manner from a fixed starting point on the plasmid. Genes located near the starting point are transferred earlier during mating than those located close to the terminus. Because mating takes time, it is very rare that a whole set of genes is transferred to a recipient; thus, those genes closest to the starting point in each strain are most often represented in the recombinants.

This experiment is based on the transfer pattern of a particularly well-studied strain of *Escherichia coli*: Hfr (ATCC 25257), a streptomycin-sensitive **prototroph** capable of growth in a mineral salts medium with glucose as the energy source (minimal medium). The recipient (F⁻) strain (ATCC 25250) is streptomycin resistant but is an auxotroph that requires leucine, proline, adenine, and tryptophan for its growth and reproduction. The media in the experiment are designed to verify parental characteristics and aid in the enumeration of recombinants.

Media 1–4 (Controls for Parental Characteristics)

1. Minimal mineral salts and glucose
2. Medium 1 + streptomycin
3. Medium 1 + leucine, proline, adenine, and tryptophan
4. Medium 3 + streptomycin

Media 5–8 (Selective for Recombinants)

5. Medium 4 less leucine
6. Medium 4 less proline
7. Medium 4 less adenine
8. Medium 4 less tryptophan

PROCEDURE (Figure 51-1)

1. You will be provided with exponential-phase cultures of *E. coli* Hfr (ATCC 25257) and F⁻ (ATCC 25250) growing in TYG broth. Two hours before the experiment, both strains will be inoculated into fresh TYG broth and incubated in a 250 ml flask in a shaking water bath. The inoculum for the F⁻ (ATCC 25250) is 2 ml into 18 ml of TYG broth. The inoculum for the Hfr (ATCC 25257) is 1 ml into 19 ml of TYG broth. After incubation for 2 hours, the Hfr culture will contain approximately 1×10^8 bacteria/ml (Klett reading at 20) and the F⁻ culture will contain approximately 2×10^8 bacteria/ml (Klett reading at 40).

2. Both cultures should be streaked on media 1, 2, 3, and 4.

3. Pipet 9 ml of the exponential-phase F⁻ culture into a sterile 250 ml Erlenmeyer flask. Add 1 ml of the exponential Hfr culture to the same flask to initiate mating. Swirl the flask to mix the 2 cultures and then let the flask stand in an incubator or water bath at 37°C. Time = 0 minutes. Prewarm a 1 liter flask containing 100 ml of minimal media to 37°C.

4. After 5 minutes (*T* = 5), carefully, without disturbing the culture, withdraw 0.1 ml of the mating mixture by pipet and inoculate it into the 100 ml of minimal media. Gently

Figure 51-1 Flow chart for demonstrating bacterial conjugation in *E. coli*.

swirl the mating mixture to produce an homogeneous suspension of mating couples. Withdraw 3 ml of the diluted mating mixture by pipet and transfer it to an empty sterile test tube. Violently vortex the test tube for 10 seconds to separate the mating pairs, which will now be too dilute to recouple. Leave the remaining dilute mating mix in the 1 liter flask undisturbed and replace it in the 37°C water bath or incubator.

5. By pipet, take 0.1 ml aliquots of the vortexed mating mix and inoculate two sets of the plates containing media 5, 6, 7, and 8. Spread the inoculum with a sterilized glass rod.

6. After the mating mix in the 1 liter flask has incubated at 37°C for an additional 10, 25, and 45 minutes ($T = 15$, 30, and 45 minutes, respectively), withdraw by pipet 3 ml aliquots and transfer them to sterile test tubes. As before (steps 4 and 5), violently vortex the text tubes for 10 seconds and inoculate 0.1 ml aliquots onto plates containing media 5, 6, 7, and 8. Spread the inocula with a sterilized glass rod.

7. Incubate all control and experimental plates at 37°C for 48 hours. If you are unable to find the time to count colonies after 48 hours, take your plates out of the incubator and store them in a refrigerator (5°C) until you can count them.

8. You should expect that some of the parental control plates (1–4) will have too many colonies to count and others will have no growth at all. Record these results on your report sheet. Count the colonies growing on media 5–8 (recombinant plates). Which characteristic is transferred most frequently from the Hfr to the F⁻ and forms the most numerous colonies? Which is the next most numerous marker to be transferred?

QUESTIONS

1. Which of the parental strains should grow on media 1–4. Why?
2. Consult a chromosome map for *E. coli* and suggest additional markers that might have been used to improve or refine the experiment you performed.
3. If you had let the experiment go for 90 or 120 minutes, would the results of your experiments have changed? How? Why?
4. On step 3, why are there ten-fold differences in the volumes of the F⁻ and Hfr cultures used in the mating mixture?

Bacterial Conjugation

Fill in the number of colonies:

Medium	Time (minutes)					Rank Order of Transfer
	0		15	30	45	
	♀	♂				
1. Control, minimal						
2. Control, 1 + streptomycin						
3. Control, complete						
4. Control, 3 + streptomycin						
5. 4 − leucine						
6. 4 − proline						
7. 4 − adenine						
8. 4 − tryptophan						

Construct a conceptual model or map of the markers on the chromosome of the Hfr strain you worked with.

52. *Bacterial Transformation*

Certain bacteria are able to transport extracellular DNA across the cell membrane into the cytoplasm. If the DNA that is taken up is derived from lysed cells of the same or closely related species, this DNA can be recombined with chromosomal DNA, which is also located in the cytoplasm. Cells that are capable of taking up high-molecular-weight DNA in such a manner are said to be **competent**. When the DNA is derived from a donor strain that has a phenotypic characteristic different from that of the recipient strain, genetic recombination can take place in such a way that some of the cells of the recipient strain will heritably carry the genes formerly found in the donor strain. The process of taking up extracellular DNA and incorporating some of it into the chromosome of the recipient strain is called bacterial **transformation**.

In this exercise you will prepare bacterial extract of a streptomycin-resistant (Str^r) culture of *Acinetobacter calcoaceticus*. This extract, which contains DNA fragments bearing the Str^r determinant, will be used to transform the competent streptomycin-sensitive (Str^s) strain for the ability to grow in media containing streptomycin.

These bacteria are aerobic, gram-negative coccobacilli usually found in soil and water but also known to reside as commensals in the human eye, ear, respiratory tract, and vagina. It has been demonstrated that *Acinetobacter* can be the causative agents of human diseases such as septicemia, meningitis, endocarditis, pneumonia, and genito-urinary tract infections, particularly in debilitated individuals.

PROCEDURE

First Laboratory Period

Prepare DNA and transformation mixture (Figure 52-1).

1. You will be supplied with a heart infusion agar plate of which one half has been streaked with the wild type Str^s strain of *Acinetobacter* (ATCC33305) and the other half with a stable Str^r (ATCC 33969) mutant derived spontaneously from the Str^s strain. Using a sterile loop, transfer a small but clearly visible amount of the Str^r cell paste to 0.5 ml of a sterile solution of 0.5% sodium dodecyl sulfate (a detergent) in "standard saline citrate" (0.15 M Na$_2$ citrate). Suspend the cell paste as completely as possible by gentle stirring by hand or with an orbital mixer. Try to avoid adherence of particles of cell paste to portions of the test tube above the liquid level. Place the covered test tube containing the suspended cells in a deep 60°C water bath and incubate for 15–60 minutes. This procedure lyses the cells and results in a sterile transforming DNA preparation.

2. With a wax marking pencil, divide the bottom of a fresh sterile heart infusion agar plate into quadrants. Label the quadrants of this plate as follows: (1) Str^s, (2) Str^r, (3) Str^s+ DNA, (4) DNA control. Using a sterile loop, transfer small amounts of cell paste from each of the cultures supplied to the center of the appropriate first two sectors of the sterile plate and gently spread this material so that it covers an area approximately 1 cm in diameter. Be certain that you sterilize the loop prior to each sampling of the cultures.

3. To the center of the third sector of the plate, add a small amount of the cells of the Str^s strain, an amount just visible to the naked eye. Use a sterile loopful of the crude DNA preparation described in Step 1 to suspend and spread this cell paste of the Str^r strain on an area approximately 1 cm in diameter.

4. Using a sterile loop, remove a loopful of the crude DNA preparation and spread it in the center of the fourth (DNA Control) sector of your plate, so that the area covered is approximately 1 cm in diameter.

5. Write your name on the bottom of the inoculated plate and place the plate in the box provided. All plates in this box will be incubated for you at room temperature until the next laboratory session.

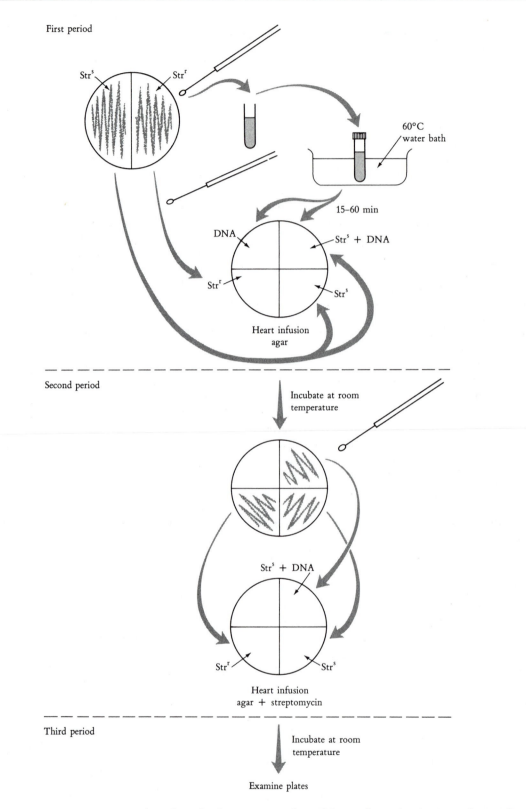

Figure 52-1 Flow chart for demonstrating bacterial transformation in *Acinetobacter calcoacedticus.*

Second Laboratory Period

Streak the transformation mixture (Figure 52-1).

1. Observe the plate you inoculated during the previous laboratory session. Bacterial cell growth should be visible in the first three sectors. There should no growth in the fourth sector, on which the DNA preparation was spread. Bacterial growth on the DNA control sector of your plate indicates that your DNA preparation was not sterile. Should this be the case, borrow a neighbor's plate when the neighbor is through with it; there will be enough material there for more than one individual. If you were reasonably careful, your DNA control should show no bacterial growth.

2. Divide the bottom of a sterile heart infusion plate containing dihydrostreptomycin (100 mg/ml) into three sectors. Label these sectors as follows: (1) Str^s, (2) Str^r, and (3) Str^s + DNA. Using a sterile loop, streak a loopful of each of the growth areas of the heart infusion agar plate to the corresponding area on the heart infusion plate containing streptomycin.

3. Write your name on the cover of the inoculated plate and place the plate in the box provided for room-temperature incubation.

Third Laboratory Period

Observe transformation.

Examine the heart infusion agar plate containing streptomycin that you streaked during the previous session. Consult the laboratory instructor if you cannot interpret your results. Discuss your results on the report sheet.

QUESTIONS

1. Why is it important that the DNA control sector of the plate observed during the second session show no growth?

2. How could you demonstrate that it is DNA in the lysed bacterial suspension that is responsible for transformation and not some other component of this crude extract?

Bacterial Transformation

Discuss the design of the experiment (including controls). Does it really prove that Str^s cells were transformed into Str^r cells by an extract from dead Str^r cells? Does it prove that DNA was the transforming factor? How could you modify the experiment to prove this? Discuss.

53. *Specialized Transduction*

Transduction involves the transfer of bacterial DNA from one bacterium to another through the agency of a **temperate** or defective **phage**. We recognize two forms of transduction: general and specialized. Almost any **genetic marker** can be transferred from donor to recipient in **generalized transduction**. When a population of sensitive bacteria is infected with a phage, some pieces of the host DNA may be incorporated by the phages as they are replicated. Lysates released from infected cells contain mixtures of normal viruses and a small proportion of viruses that have incorporated some of the host DNA. If the virus can form a temperate relationship with a new host, the old host's genes may be recombined with the new host's DNA. **Specialized transduction** involves the transfer of a specific group of host genes by a lysogenic phage which becomes integrated into the host DNA at a specific site. In this experiment, we will use the well-known lambda phage of *E. coli* to demonstrate specialized transduction at the galactose locus.

PROCEDURE

A. Induction of Lysis-Virus Propagation

1. You will be given an exponential-phase culture of *E. coli* (ATCC 25256; galactose + bacterial phage lambda) grown on TYG broth. Warm a 10-ml tube of TYG broth (37°C). When the cells have reached a density of approximately 5×10^8 cells (40 Klett units), aseptically transfer 10 ml of the culture to a sterile empty petri dish.
2. Remove the petri dish lid and expose the culture to a dose of UV light which your instructor has previously determined will kill 10% of the *E. coli*. (With the Gates Raymaster desk-type germicidal lamp, Type A, 6281-K25 at 8 cm, this dose is 20–40 seconds.)
3. Without removing the lid treat a control culture in a similar manner.
4. Immediately after exposure aseptically and separately remove each of the cultures (experimental and control) by sterile pipet and transfer each to the 10 ml prewarmed tube of TYG broth.
5. Incubate the tubes in a shaking water bath for 3–4 hours. Then observe lysis of the UV-exposed cultures. Add 0.5 ml of chloroform to the 20 ml lysed culture, shake it briefly, and store it at 5°C in an **explosion-proof refrigerator** until the next period.

B. Transduction (Figure 53-1)

1. With a pipet, carefully transfer 0.1 of the culture lysate prepared in part A, step 5 to two minimal galactose agar plates. Aseptically spread them with alcohol-sterilized glass rods. Incubate the plates at 37°C for 5 minutes. Leave the plates partially uncovered during incubation to make sure that any transferred chloroform has evaporated.
2. Label one of the two plates *lysate sterility control* and return it to the incubator. To the other plate, aseptically pipet a 0.1 ml aliquot of an overnight culture of *E. coli* (ATCC 27161, Gal (−) grown in TYG broth. Aseptically spread the *E. coli* with a sterile glass rod. Label this plate Gal (−) *experimental.*
3. Take an 0.1 ml aliquot of the same Gal (−) culture (ATCC 27161) and inoculate it onto a third minimal galactose agar plate. Label this plate Gal(−) *control.*
4. Incubate all three plates at 37°C for 48 hours and then record your observations.

Demonstration of Viruses in the Lysate and Their Enumeration

1. Prepare serial 10-fold dilutions of the virus lysate (from part A, step 5) by pipetting 1 ml of lysate into 9 ml of TYG broth. Continue the 1:10 dilutions until seven serial dilutions are prepared.
2. In duplicate, pipet 0.1 ml samples from dilution tubes 5, 6, and 7 into molten tubes of top-layer agar (3.5 ml each). By pipet, add 0.2 ml of the overnight culture of the Gal (−) *E. coli* culture (ATCC 27161) to each of

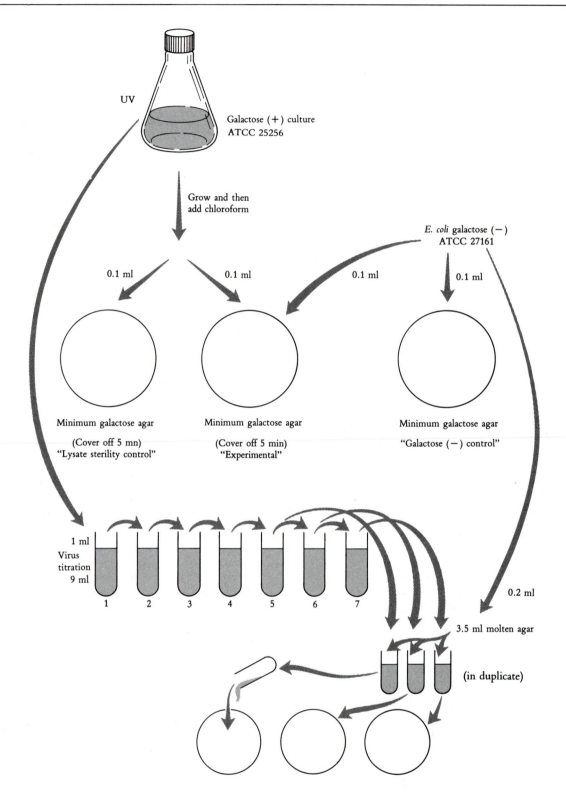

Figure 53-1   Flow chart for demonstrating specialized transduction by the lambda phage and measurement of lambda titer.

the six tubes. Pour the contents of all six tubes onto thick, freshly made TYG agar plates. Spread by tilting the plates back and forth until the layer is even. Incubate at 37°C overnight.

3. Prepare a control plate (no viruses) by inoculating only the Gal (−) *E. coli* (ATCC 27161) in the top-layer agar. Spread the top-layer agar on a TYG agar plate. Incubate at 37°C overnight. Count the plaques on the

plates in which they are clearly separated from each other (~30–300).

4. Discuss the experiment on your report sheet.

QUESTIONS

1. Explain the purpose of the various controls in the experiment.
2. Why is it important to make sure that the chloroform has evaporated in part B, step 1?

Specialized Transduction

Discuss the design of the transduction experiment. Don't forget the controls. Did you prove that it was a virus that changed a Gal (−) strain to a Gal (+) strain?

E L E V E N

Viruses

The viruses are ultramicroscopic—too small to be viewed with the light microscope, visible only with the greater resolution of the electron microscope. They are *particulate*, not cellular, being more or less macromolecules composed primarily of a nucleic acid genome, either DNA or RNA, and protein. They are obligate parasites whose nucleic acid genomes control and utilize the synthetic capacities of their host cells for replication.

Basically, the viral particle, or **virion**, is a nucleic acid core surrounded by a protein coat, or **capsid** (Figure 56-1), composed of protein subunits or **capsomers**. In some more complex viruses, the **nucleocapsid** (core plus coat) is surrounded by an additional envelope, and some have spikelike surface appendages.

Morphologically, the viruses can be divided into three groups: rod-shaped virions, cubical or polyhedral virions, and those possessing more complex structures.

The viruses would have gone unnoticed during the early development of microbiology except for the fact that they are infectious agents, evident from the symptoms of the diseases they caused. Even today, when we can "see" a virus with the electron microscope, we still depend for study primarily on the symptoms of experimental virus infections. Thus, for studying a bacterial virus, or **bacteriophage** (bacteria-devouring), we observe the infection of laboratory cultures and, for studying plant viruses, the experimental infection of plants. For studying animal viruses, **chick-embryo culture** and **cell cultures** have replaced the experimental infection of animals.

54. *Isolating Bacteriophages and Observing Their Characteristics*

Bacterial viruses, commonly called *bacteriophages*, can be isolated from many environments. They cannot live independently, so they are sought where their host cells are found. If you were interested in isolating a phage for enteric organisms, raw sewage would be a source. Soil would be a source of phages for spore formers and other soil-dwelling bacteria.

In ordinary conditions, the numbers of phage in any natural source are not high, and one of the first steps in the recovery of phage is an enrichment procedure in which the phage source is incubated with a culture of host cells to increase the phage. As the bacterial cells increase, viral particles are formed; the cells eventually disrupt and release the virions into the medium. Following this enrichment procedure, the sample containing the increased phage population is centrifuged to remove coarse material and then passed through a bacterial filter to remove bacterial contaminants. The clear liquid should contain phage. You can demonstrate this by mixing some of the liquid with a young culture of uninfected host cells and layering the mixture on the surface of an agar plate. After incubation, the film of bacterial growth will be mottled by clear circular areas called **plaques** (Figure 54-1). These plaques represent areas of phage reproduction and lysis of the population of infected bacterial cells in the area.

In this exercise you will attempt to isolate a bacterial virus for *Escherichia coli* because phages for this bacterium are relatively common.

PROCEDURE

Students should work in pairs.

1. Add 45 ml of raw sewage to 5 ml of concentrated broth provided. (*Note*: In this deca strength broth, ingredients are at about ten times the usual concentration to provide the proper level of nutrients after dilution with the sewage.)
2. Inoculate the mixture with 5 ml of a 24-hour culture of *E. coli* and incubate at 37°C for 24 hours. This step is an enrichment procedure for increasing the number of phage.

3. Clarify 10 ml of the sewage culture by centrifuging at 2500 rpm for 10 minutes.
4. Pass the centrifuged liquid through a bacteriological filter (this removes organisms that would overgrow the plates as the exercise continues). The clear liquid can now be tested for phage.
5. Melt three tubes of regular nutrient agar (12 ml/tube) and three tubes of soft nutrient agar (0.7% agar/3 ml tube). Temper the agar to 45°C and pour one tube of the regular nutrient agar into each of three petri plates and allow to harden.
6. After tempering the three tubes of soft agar (3ml/tube) to 45°C, add 0.1 ml of a 24-hour *E. coli* culture (the same strain used in Step 2) to each tube.
7. To the first tube of inoculated soft agar, add one drop of the sewage-culture filtrate (from Step 4), mix, and pour onto the surface of one of the nutrient-agar plates prepared in step 5. To a second tube of inoculated soft agar, add 5 drops of the filtrate from Step 4, mix, and pour onto the surface of a second agar plate from Step 5. As a control, pour the third tube of the inoculated soft agar on the other agar plate from Step 5. Allow the soft agar to solidify and incubate the plates (inverted) at 37°C until plaques are visible (6 – 24 hours).
8. Examine the plates that you seeded with the sewage cultures for plaques. These are areas in which phage particles have lysed young growing cells of *E. coli*.
9. Isolate a pure culture of phage by cutting out a single plaque with a sterile loop and transfer it to a young culture of the original host *E. coli*.
10. Incubate the infected *E. coli* and an uninfected control at 37°C with gentle shaking for 3 hours. Observe for clearing (lysis) of the infected tube.
11. Record and illustrate the results on your report sheet.

Bacteriophage, like all viruses, is specific for its host. The bacterial virus you have isolated in this exercise is specific for the strain of *E. coli* (or one or two closely related strains) that you employed in the enrichment procedure of Step 2. You can try to demonstrate this by infecting young broth cultures of several different strains of *E. coli* (which your instructor can provide) with the phage that you isolated.

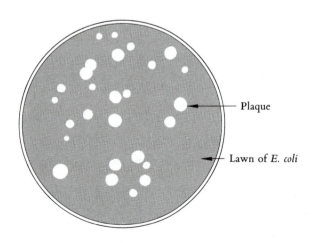

Figure 54-1 Bacteriophages on a lawn of *E. coli*.

QUESTIONS

1. Why doesn't the phage plaque keep spreading until the whole plate is cleared?
2. Why should all glassware, media, and equipment used in this exercise be steamed or autoclaved following use?
3. How could you titer (that is, count) the number of phage in the original suspension?
4. Can you use the techniques of this experiment to demonstrate temperate phage?

Isolating Bacteriophages and Observing
Their Characteristics

Name _____

Desk No. _____

Indicate the number and appearance of your phage plaques.

How long did it take for the phage-infected young culture of *E. coli* to clear?

55. *Bacteriophage Production: Single-Step Growth*

The reproductive cycle of a bacteriophage begins with its adsorption to a sensitive host cell. **Adsorption** is the attachment of phages, some of which are tadpole-shaped, tail-first to the cell wall of a host. The tail then functions somewhat like a hypodermic needle, enzymatically piercing the wall of the host cell and injecting the nucleic acid core of the phage (Figure 55-1).

Inside the cell this nucleic acid loses any visible resemblance to the mature phage and is referred to as the **vegetative phage**. During this phase, called the **latent** period of the **lytic**, or reproductive, cycle of a virus, the phage takes over the synthetic capacity of the host cell and rapidly replicates itself.

Protein is synthesized to encase each prophage, and then the host cell lyses, releasing up to several hundred daughter phages. The number of phage particles released per cell is called the **burst size**.

Unlike the reproduction of bacteria, the reproduction of bacterial viruses appears to us as a one-step procedure, with no increase in numbers until the end of the latent period, when the cell lyses and a sharp rise in the number of phage particles occurs.

In this experiment the lytic cycle is followed with the one-step growth technique, in which a suspension of sensitive bacteria is mixed with a phage suspension, and the mixture is incubated until adsorption occurs. The mixture is then di-

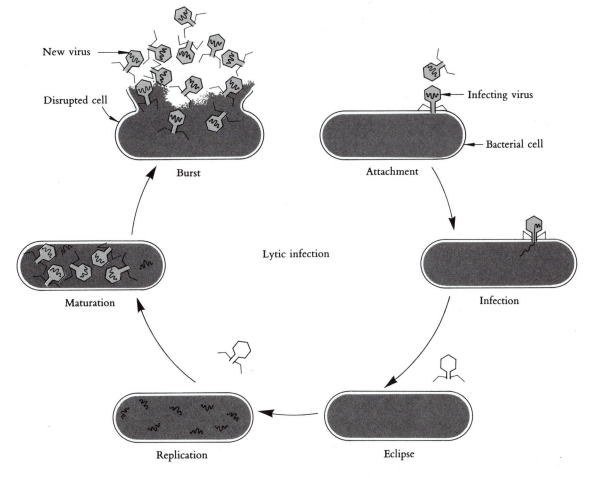

Figure 55-1 The lytic cycle.

luted, preventing further adsorption. Samples are withdrawn and plaque counts made at frequent intervals. With this information, and knowing the original bacterial count as well as the degree of dilution, you can calculate the average number of phage particles per bacterial cell at each time interval. A plot of this phage count versus time is an expression of the single-step growth response of a phage infection. From this curve you can also calculate the latent time and the average burst size of the infection.

PROCEDURE

Students should work in groups of two or more.

1. Plate out an *Escherichia coli* strain B culture at dilutions of 10^{-5}, 10^{-6}, and 10^{-7}.
2. Pipet 0.9 ml of *E. coli* into a sterile test tube.
3. Pipet 0.1 ml of a T2 bacteriophage suspension into the 0.9 ml of *E. coli*. Let stand for 5 minutes. During this period, phage particles are absorbed by approximately 80% of the bacterial cells. *Note*: The timing of this and the following steps is critical.
4. Aseptically pipet the phage and *E. coli* mixture into a sterile centrifuge tube and centrifuge. Decant the supernatant, resuspend the pellet of cells in 0.9 ml of nutrient broth, and mix well by swirling. These steps remove unadsorbed phage particles.

5. Add 0.1 ml of the cell suspension from Step 4 to 9.9 ml of nutrient broth and mix well by swirling. (What dilution of the original cell culture is this?)
6. Add 0.1 ml of the diluted cell suspension from Step 5 to 9.9 ml of nutrient broth. Mix well by swirling. Incubate at 37°C in the water bath provided.
7. Melt five tubes of regular nutrient agar and five tubes of soft nutrient agar (0.7% agar). Temper the agar to 45–50°C and pour one tube of the regular nutrient agar into each of five petri plates, and allow to harden.
8. After tempering the five tubes of soft agar to 45°C, add 3 or 4 drops of the original *E. coli* strain B culture to each tube. Considering Step 3 to be T=0, add a 0.1 ml sample from the diluted culture of Step 6 to a different tube of the inoculated soft agar at intervals of 20, 25, 30, 35, and 40 minutes. Mix each tube immediately after it is treated, pour it on the surface of one of the nutrient agar plates prepared in Step 7, and record the time of its treatment. Allow the soft agar to solidify and incubate the plates (inverted) at 37°C until the following laboratory period.
9. Determine the count of *E. coli* from the plates made in Step 1. Examine the plates from Step 8 and count and record on your report sheet the numbers of plaques formed at the various time intervals.

REPORT **55**

Bacteriophage Production:
Single-Step Growth

The bacterial count of your original sample was _____ /ml. The bacterial count after Step 6 was _____ /ml (calculate from dilutions).

Record the number of plaques per plate (from Step 9) and complete the table. Phage particles per cell = number of plaques ÷ number of bacteria (count at Step 6).

Time (minutes)	Plaques per plate	Phage particles per cell
20		
25		
30		
35		
40		

Plot the increase in phage against time.

56. *Isolating Tobacco Mosaic Virus and Infecting Plants*

The tobacco mosaic virus (TMV) was the first to be discovered, the first to be purified, and the first to be crystallized. This rod-shaped or helical virus (Figure 56-1) infects its host tobacco plant via a wound. Depending on the strain of tobacco, the infection of TMV manifests itself either as a systemic infection in which there is a general wrinkling and shriveling of all the plant's leaves or as a local infection in which only the infected leaf shows necrotic lesions, which appear as small brownish plaques. Symptoms of a systemic infection take 1–3 weeks to appear; a localized infection is evident in 2–4 days.

In this experiment you will first isolate TMV from an infected tobacco plant by grinding up diseased leaves and filtering the exudate. The filtrate will then be used to infect two types of tobacco plants, designated NN and Turkish. Because these are healthy plants, you will need to wound the leaf cells by using an abrasive carborundum before inoculating the plants with the virus (Figure 56-2).

PROCEDURE

Work in groups of 5 or 6. *Note:* Wash hands carefully with soap and water before and after you inoculate healthy plants so that you will not spread the infection.

1. Each group of students is given a diseased tobacco plant. Remove the infected leaves from the plant and grind them up in a sterile Waring blender containing 100–150 ml of sterile water.

2. Filter out the particulate matter by pouring the contents of the blender through a funnel, in which you have placed filter paper, into a flask.

3. Refilter the filtrate from Step 2 through a bacteriological filter (0.45 μm). This filtrate should contain only TMV in suspension.

4. Each group of students is given an NN and a Turkish tobacco plant. Place a very small piece of tape on the ends of two or three leaves on each plant. These are the leaves you will inoculate.

5. Spray a small amount of carborundum from a wash bottle onto one of the leaves to be inoculated.

6. Gently rub the leaf with a piece of cotton that has been soaked in the filtrate from Step 3. Be careful to abrade the leaf only slightly.

7. Wash the carborundum and excess filtrate from the leaf with a wash bottle of distilled water.

8. Repeat Steps 5–7 on each leaf to be inoculated. Your instructor will maintain a set of control plants for your reference.

9. Check the plants during each laboratory period for up to 3 weeks. Look for both localized and systemic symptoms (Figure 56-3).

10. Describe and illustrate the symptoms on your report sheet.

Some commercial tobacco products, particularly foreign brands, carry tobacco mosaic virus. You may be interested in attempting to isolate tobacco mosaic virus by treating a plant that is subject to localized infection with the tobacco from various

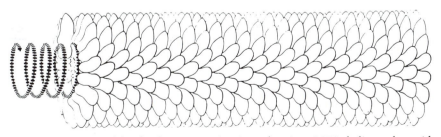

Figure 56-1 Model of tobacco mosaic virus showing RNA helix and capsid composed of capsomeres. (After H. Fraenkel-Conrat, "The genetic code of a virus," *Scientific American*, October 1964.

Remove infected tobacco leaves

Grind up the leaves

Filter out the leaf particles

Filter out the bacteria

Infect a tobacco plant

CARBORUNDUM

Figure 56-2 Flow chart illustrating steps in the infection with tobacco mosaic virus.

Uninfected tobacco plant   Systemic infection   Localized infection

Figure 56-3   Tobacco mosaic virus infection of tobacco plants.

types of cigarettes, pipe tobacco, and so on. Russian and Turkish brands are often good sources of these viruses.

QUESTIONS

1. What is the historical importance of TMV?
2. In your plant with a systemic infection, which leaves show the symptoms first?
3. How are most plant viruses transmitted?
4. How could a quantitative assay of infective plant viruses be developed?

REPORT 56

Isolating Tobacco Mosaic Virus
and Infecting Plants

Name _____

Desk No. _____

Draw the appearance of a localized infection on a leaf.

Describe the appearance of a tobacco plant with a systemic infection.

T W E L V E

Water and Air Microbiology

Because of our increasing population and developing technologies, the earth's supply of potable water has become a precious resource to be carefully guarded. Everyone today must become aware of the pressing need for clean water.

Typhoid epidemics and dead fish on the shores of a lake are both symptoms of the pollution of water resources. Dumping wastes into a body of water causes radical and detrimental changes in the types, numbers, and activities of its microbes. Instances of the spread of diseases such as typhoid fever and dysentery by fecal pollution of water supplies are familiar, but water as a source of disease is only one problem that results from water pollution. As organic wastes enter a stream or a lake, microbial degradation of the aquatic environment depletes the oxygen supply, creating an environment that is devoid of all but anaerobic forms of life, an environment characterized by dead fish, decaying plant life, and the stench of the activities of anaerobic microbes. Of course, some industrial wastes of other types cannot be degraded by microbes, so these exert their own destructive effects on aquatic life.

To control our misuse of our waters, we must have methods of detecting and measuring pollution. The quality of water is determined largely by bacteriological analyses. The important objective of these analyses is to determine whether a water supply contains fecal organisms (not necessarily pathogens) whose presence is indicative of pollution by human or animal wastes. The organisms sought are usually the **coliform bacteria**, or the coli–aerogenes group, which includes all aerobic and facultative gram-negative, non-spore-forming straight rods that ferment lactose with gas formation. These are the *indicator* organisms of water quality.

The ideal bacterial indicator of pollution must be applicable to all types of water. Indicator organisms must always be present in water when bacterial fecal pathogens are present and should have a population density directly related to the degree of fecal pollution. Indicator organisms should survive longer in the water than pathogens and should disappear rapidly after the pathogens disappear. They should be absent from safe water. The best indicator organisms should be easily quantitated without interference by other bacteria and should be relatively harmless to the technician. Coliforms fulfill most of these requirements with the exception that some do not disappear after the pathogens have died.

Some coliforms are found in nonfecal sources, but the presence of fecal coliforms (FC) specifically indicates recent fecal waste from warm-blooded animals. They are identified by their ability to ferment lactose at 44.5°C with the production of acid and gas. These are typically *Escherichia coli* of the IMViC (indole, methyl red, Voges–Proskauer, citrate) type $++--$ as well as *Enterobacter aerogenes* ($--++$) with some intermediate forms. The fecal streptococci (FS) are more numerous inhabitants of the animal intestine (rather than the human intestine) and the FC/FS ratio, that is, the relative proportions of each in a sample, is effective in estimating the likelihood of an animal source of fecal pollution.

The amount of organic materials in water or waste can also be ascertained by measuring the oxygen consumed by microorganisms, which is proportional to the organic material present. This principle is used in the **biochemical oxygen demand** (B.O.D.) technique. This test, in addition to being a measure of organic wastes in a water supply, can determine the effectiveness of a sewage treatment system; it can also be used to determine whether waste is sufficiently stabilized to be disposed of in a lake, river, or stream without endangering the oxygen level.

57. *Analysis of Water*

The standard analysis of water for the presence of coliform bacteria is in three parts: the **presumptive** (Figure 57-1) the **confirmed**, and the **completed tests** (Figure 57-2). In the presumptive test you look for microorganisms, presumably coliforms, that ferment lactose with gas production. In the confirmed test, you transfer cultures that show growth and gas production in the presumptive test onto media that are selective and differential for the coli–aerogenes group. In the completed test, you isolate and grow pure cultures of the organisms that gave reactions typical of coliforms on the confirmatory media. This analytical scheme provides a sensitive procedure for detecting these *indicator* organisms.

PROCEDURE

For this exercise you may want to bring a sample of water from your home water supply, a nearby stream or pond, or some other source for part of the analysis. *Note*: Full details of a somewhat different procedure can be found in *Methods for the Examination of Water and Wastewater*, 17th edition (1989), published by the American Public Health Association, 1015 Eighteenth St. N.W., Washington, DC 20036.

A quantitative methodology (MPN) for the presumptive test was given in Exercise 18 (also see Table 57-1).

Figure 57-1 Presumptive test for the coliform group, showing gas production from lactose broth.

Presumptive Test

1. Inoculate 10-ml portions of each water sample provided into a different one of three large tubes containing 10 ml of lactose broth that has been prepared with twice the normal concentration of constituents to allow for the dilution. Inoculate 1.0 ml and 0.1 ml of water into small tubes (two sets of three each) of single-strength lactose broth.
2. Incubate at 37°C for 2 days.
3. Observe after 24 and 48 hours. The presence of gas in any tube after 24 hours is a *positive* presumptive test. The formation of gas during a second 24-hour period is a *doubtful* test. The absence of gas formation after 48 hours incubation constitutes a *negative* test, indicating that the water supply does not contain coliforms.

Table 57-1 Most Probable Numbers (MPN) of Bacteria per 100 ml of Sample, Using Three Tubes of Each Dilution

Number of Positive Tubes in Dilutions			MPN per 100 ml	Number of Positive Tubes in Dilutions			MPN per 100 ml
10 ml	*1 ml*	*0.1 ml*		*10 ml*	*1 ml*	*0.1 ml*	
0	0	0	3	2	0	0	9.1
0	1	0	3	2	0	1	14
0	0	2	6	2	0	2	20
0	0	3	9	2	0	3	26
0	1	0	3	2	1	0	15
0	1	1	6.1	2	1	1	20
0	1	2	9.2	2	1	2	27
0	1	3	12	2	1	3	34
0	2	0	6.2	2	2	0	21
0	2	1	9.3	2	2	1	28
0	2	2	12	2	2	2	35
0	2	3	16	2	2	3	42
0	3	0	9.4	2	3	0	29
0	3	1	13	2	3	1	36
0	3	2	16	2	3	2	44
0	3	3	19	2	3	3	53
1	0	0	3.6	3	0	0	23
1	0	1	7.2	3	0	1	39
1	0	2	11	3	0	2	64
1	0	3	15	3	0	3	95
1	1	0	7.3	3	1	0	43
1	1	1	11	3	1	1	75
1	1	2	15	3	1	2	120
1	1	3	19	3	1	3	160
1	2	0	11	3	2	0	93
1	2	1	15	3	2	1	150
1	2	2	20	3	2	2	210
1	2	3	24	3	2	3	290
1	3	0	16	3	3	0	240
1	3	1	20	3	3	1	460
1	3	2	24	3	3	2	1100
1	3	3	29				

Figure 57–2 Flow diagram for the analysis of water.

Confirmed Test

This test should be applied to all samples that give a positive or doubtful presumptive test.

1. From the tube of lactose broth testing positive with the smallest inoculum of water, streak a plate of eosin–methylene-blue (EMB) agar or Endo agar to give well-isolated colonies.
2. Incubate at 37°C for 2 days.† If typical colonies (as described in Exercise 32) have developed on the plate within this period, the confirmed test can be considered positive.

Completed Test

1. From the EMB or Endo plate, pick two colonies that you consider most likely to be organisms of the coliform group, and transfer one to an agar slope and the other to a lactose-broth fermentation tube. Coliform organisms on Endo or EMB agar form darkish colonies that often, but not always, exhibit a greenish metallic sheen. Both types should be carried into the completed test if they appear.
2. Incubate at 37°C for 2 days.
3. From the agar slope, make a gram stain and a spore stain.
4. Discuss the results of your analysis on the report sheet.

The formation of gas in the lactose broth and the demonstration of gram-negative, non-spore-forming rods in the agar culture constitute a positive test, revealing the presence of coliform group bacteria and indicating that the water sample was polluted. (See Figure 57-2 for a flow diagram of the standard analysis of water.)

QUESTIONS

1. What microorganisms other than coliforms are likely to give a positive presumptive test?
2. Why not test water samples directly for *Salmonella typhosa* or other pathogens?

---

† *Standard Methods for the Examination of Water and Wastewater* recommends an incubation temperature of 35°C throughout the analysis: 37°C can be used if it is more convenient, without significantly altering the results.

REPORT **57**
Analysis of Water

Name _____

Desk No. _____

Consider the results of your analysis in terms of public health. Would you (1) drink your sample, (2) wash food in your sample, (3) swim in your sample?

Make a list of authorities or public laboratories who carry out standard analyses of water on a regular schedule to protect public health.

58. *Membrane Filter Technique in Water and Air Analysis**

WATER ANALYSIS

A membrane filter has been developed that filters bacteria from water and other materials. The trapped bacteria are then grown directly on the filter by placing it on a suitable growth medium. The use of a selective medium determines the numbers of coliforms and certain other species of bacteria in the sample. This technique also offers promise for the direct isolation of some organisms, such as *Salmonella*, even when they are present only in small numbers.

In principle, a measured amount of sample is filtered through a membrane with a pore size of about 0.45 μm, which traps the bacteria on its surface. The membrane is then placed on a thin absorbent pad that has been saturated with a medium designed to grow or permit differentiation of the organisms sought—a modified Endo medium, for example, if total coliform organisms are sought. Fecal streptococci may be selected by use of a modified *Enterococcus*-agar medium (KF agar) containing azide, and fecal coliforms can be selected by using a modified fecal coliform medium (M-FC broth) with incubation at 44.5°C. After incubation in a small petri plate, the colonies are counted under low magnification.

The success of this method depends on using effective differential or selective media that will enable easy identification of colonies. This method has advantages over the traditional water analysis procedure in Exercise 57 because it is more direct and quicker (giving results in 18–24 hours) and can easily test large volumes of water (hence yielding more accurate results).

*Equipment and materials for the examination of water and air by membrane filter techniques are available from the Millipore Corp., Bedford, MA 01730. Further details of the procedures are found in the Millipore Corp. publications numbered LAM 3020/U and LAP 3090/U, and in the publication of the American Public Health Association referred to in Exercise 57.

PROCEDURE

Total Coliform Test

1. Attach the filter holder to the rubber stopper, insert into the vacuum flask, and connect the flask to the vacuum line or aspirator (Figure 58-1).
2. Using sterile forceps, transfer a sterile membrane filter (Millipore HAW 604750) to the platform base of the filter unit. Place the filter with ruled side up.
3. Place the matched funnel unit over the filter disc, making sure that it is clamped in place firmly by the scissors-type clamp.
4. Line up three small petri dishes, labeled with the three sample volumes to be used and your initials.
5. Pour about 20 ml of sterile buffered water into the funnel before adding your sample. Your instructor will guide you on the volume of water to use. In this exercise, you can test water from a stream, river, pond, sea beach, or any body of water that interests you. Your instructor will help you make a decision or provide you with a test sample.
6. Shake your bottle of sample vigorously.
7. Measure out the volume of sample (1, 4 and 15 ml) to be tested with a 10 ml pipet or a graduated cylinder and introduce it into the funnel.
8. Rinse the measuring vessel twice with an amount of sterile buffered water that is equal in volume to the sample and add this water to the funnel (this rinses cells from the measuring vessel).
9. Turn on the vacuum motor or water aspirator and allow all the liquid to pass through the filter into the flask.
10. Leaving the vacuum on, rinse the funnel with a volume of sterile buffered water that equals the total amount of liquid filtered; pour this rinse water onto the inside wall of the funnel so that a swirling wash results.
11. Allow the rinse water to pass entirely through the filter and then repeat with a

Figure 58-1 A type of membrane-filter apparatus.

second equal rinse. After all the water has passed through the filter, allow the vacuum to run 1 minute or until the filter appears dry.

12. If you are using the Millipore specialized 47 mm petri dishes (PD 100 4700) and other materials in the water testing kit, prepare one of the dishes as follows:

(a) Remove a presterilized pad from the package of pads and filters (HAWG04750) with the aid of sterile forceps and place it in the petri dish (Figure 58-2A).

(b) Take an ampule of sterile Endo medium and place it in an ampule breaker (Figure 58-2B).

(c) Break the ampule and pour its contents onto the presterilized pad (Figure 58-2C).

13. Turn off the vacuum source and move the membrane filter with flamed sterile forceps to the Endo medium in a small petri dish (Figures 41-2D, E).

14. Push the membrane against the far side of the petri dish and onto the medium and roll it onto the medium to avoid trapping air bubbles under the membrane (Figure 58-2E). The medium will diffuse from the pad through the filter to support the growth of bacteria on the upper surface of the filter (Figure 58-2F).

15. Incubate at 37°C for 22–24 hours.

16. Examine the plates and note colonies that are pink to dark red with a golden green metallic sheen. Count plates that contain 20–80 such colonies of coliforms and no more than 200 colonies of all types.

17. Calculate the numbers of organisms per 100 ml using the formula:

$$\text{indicator count per 100 ml} = 100 \times \frac{\text{total number of colonies counted}}{\text{number of ml of the sample tested}}$$

Fecal Coliform Test

Repeat Steps 1–11 using 1, 3 and 10 ml volumes.

12. (a) Repeat Step 12a.
(b) Using a 10 ml pipet, add 2 ml of M−FC broth to the surface of each absorbent pad.

13. Aseptically transfer the membrane filter to the top of the absorbent pad.

14. After snapping the petri plates shut, seal them with waterproof tape, insert them into a waterproof plastic bag, and incubate them in a 44.5°C water bath for 22 hours. Be sure to sink the bags beneath the surface.

Figure 58-2 Flow chart illustrating steps in the membrane filter method of enumerating coliform bacteria in water. *A*: Place sterile absorbent pad into petri dish. *B*: Crack open sterile vial of medium. *C*: Aseptically add medium to pad. *D*: Aseptically remove filter from the filtering apparatus. *E*: Place filter on top of the absorbent pad in the petri plate. *F*: Cutaway side view of the filter on the pad showing the diffusion of medium from pad to filter and growth on filter surface.

15. Count blue-colored colonies with coliform characteristics. Use the plate containing 20–60 colonies.

Fecal Streptococcus Test

Repeat Steps 1–11 using sample volumes of 1, 5, and 25 ml; label the three plates with these volumes.

12. Repeat Steps 12a–c of the Total Coliform Test, but use KF medium.
13. Aseptically transfer the three membrane filters to the top of the absorbent pad of each plate.
14. Incubate the prepared plates for 48 hours at 37°C.
15. Examine the plates for colonies that are light pink and flat and for smooth, dark-red colonies with pink margins. Count the plate that has 20–100 colonies.

Collect the class data for fecal coliform (FC) and fecal streptococci (FS) and calculate the ratio FC/-FS:

$$\frac{\text{number of fecal coliform per ml}}{\text{number fecal streptococci per ml}}$$

An FC/FS greater than 4 shows strong evidence of pollution derived from human waste. An FC/FS ratio less than 0.7 indicates pollution derived predominantly or entirely from livestock or poultry waste. If the ratio is between 2 and 4, it suggests a predominance of human waste in mixed pollution. A ratio of 1 to 2 cannot be interpreted; it is then suggested that a sample nearer the point source of pollution be taken.

16. Record your results on the report sheet.

AIR ANALYSIS

Many organisms spread to new environments on air currents, but medically significant organisms are only a small part of the airborne load. Dispersal and survival of organisms in the air in a hospital environment can often cause hospital-acquired (nosocomial) infections. Aerosol droplets from a cough or sneeze contain millions of microorganisms and are a well-recognized source of organisms; however,

skin, hair, bed linen, clothing, brooms, and mops also contribute to the airborne load. Some organisms caught up in air have special attributes that enable them to survive; many cease normal metabolic activity until they are deposited in a suitable environment. Thus, it is important to sample the air in such an environment to monitor these airborne organisms.

An air-sampling apparatus, or impinger, has a membrane filter that collects the microorganisms and supports colony growth after it is placed on the appropriate medium. The organisms are first deposited in a special broth above the membrane. The air flow is then stopped, and the entire broth solution is drawn through the membrane, which is then removed, placed on culture medium, and incubated. Quantitation is accomplished by passing the air stream through a limiting orifice that permits a flow of 10 liters per minute.

In this exercise, you will compare the airborne cell population during the laboratory period with that sampled by your instructor before class.

PROCEDURE

1. Set the Sterifil apparatus in the area you want to sample (Figure 58-3). Attach it to the vacuum source.
2. Turn on the vacuum source. The vacuum draw produces an air current that pulls a sample of air down onto the impingement fluid. The volume of air passing through the orifice must be large enough to be measur-

Rubber sealing cap
Apply primary vacuum
Limiting orifice
Air intake
Impingement fluid
Membrane filter
Apply secondary vacuum

Figure 58-3 Sterfil apparatus for sampling air.

able and representative. In general, a sample size of 10 cubic feet (280 liters) is sufficient for most indoor sampling. Since the desired sample is 280 liters, the optimum sampling time with a 10-liter per minute orifice is 28 minutes. Use a time or stopwatch to monitor sampling time.

3. While the apparatus is gathering the sample, prepare two petri dishes. Place a sterile absorbent pad in the bottom of each petri dish and saturate the pad with 1.8–2.0 ml of the appropriate (a or b) broth-culture medium.

    Plate a, total count medium with indicator. This is a nonselective medium supplied in 2 ml ampoules for growing a wide spectrum of airborne microorganisms.

    Plate b, yeast and mold medium. This is a differential medium that is buffered at a pH between 4 and 5 to discourage the growth of other organisms such as bacteria.

4. At the end of the sampling period, turn off the vacuum system.

5. Disconnect the aerosol adapter. Lift the funnel cover and, using a squirt bottle with sterile water, rinse down the inside walls of the Sterifil funnel.

6. Remove the aerosol adapter and attach the vacuum tubing to the longer sidearm of the Sterifil receiver flask. Apply a gum rubber cap to the opening on the other sidearm.

7. Turn the vacuum on. This causes the rubber funnel outlet cap to be sucked into the receiver flask along with the impingement fluid. Thus, the apparatus need not be disassembled to remove the funnel outlet cap.

8. Turn off the vacuum and rinse the funnel walls again as in Step 5. Turn the vacuum back on to draw all the rinse through the filter.

9. Turn off the vacuum, unscrew the Sterifil funnel, and place it safely aside.

10. Ignite the sterilizing alcohol on the tips of the forceps, let cool, and gently remove the filter from the filter-holder base.

11. Touch the far edge of the filter to the far edge of the pad in the petri dish. Let the

filter roll down gradually over the pad. Avoid trapping air bubbles.

12. Seal the two halves of the petri dish tightly together. Turn the pad and filter upside down and place the culture in an incubator for 24 hours at 37°C.

13. At the end of the incubation period, observe the number of bacteria or yeast and mold colonies and calculate the number of those organisms in the air sample by using the following formulas:

[sampling time (minutes)] ×
[capacity of the limiting orifice (liters per minute)] = number of liters per sample

$$\frac{\text{liters of sample}}{28} = \text{cubic feet}$$

$$\frac{\text{number of colonies observed}}{\text{cubic feet of sample}} =$$

number of organisms per cubic foot of air

14. Record your results on the report sheet.

Making Permanent Records

In some cases you may want to preserve the results of your air sampling for future reference or to attach to a report. The technique is as follows:

1. Using forceps, remove the specimen filter carefully from the petri dish and set it on a dry blotter pad or paper towel for 30–45 minutes or until dry.

2. The dry filter can now be preserved and permanently affixed to a report using a 3″ × 3″ piece of transparent contact tape.

QUESTIONS

1. What are other applications of the membrane filter technique?

2. What is the purpose of preincubation of the membrane filter in the enrichment medium in water analysis?

Membrane Filter Technique in Water and Air Analysis

WATER ANALYSIS

What was the total coliform count of the water sample?

What was the fecal coliform count of the water sample?

What was the fecal streptococcus count of the water sample?

Based on the FC/FS ratio would you guess that the pollution of the water was of human or animal origin?

AIR ANALYSIS

What was the bacterial count of the air sample you examined?

What was the yeast and mold count of the air sample?

Show how you calculated these results.

Food Microbiology

Eating utensils and foods can be sources of disease caused by microorganisms. The growth of microorganisms in food can also bring about changes in it, undesirable, spoilage, or beneficial, producing a more easily preserved food or a food with a more desirable flavor. Although the subject of food microbiology is broad, the principles are merely applications of the basic microbiology described in earlier sections of this manual. For example, food preservation entails **asepsis** to prevent contamination and environmental extremes such as acid or heat to inhibit or destroy microorganisms. Microbial spoilage of a food is an example of the effects of selective and enrichment cultures, with the pH and the chemical composition of the food primarily determining the form of the spoilage.

The exercises in this section illustrate techniques used to determine numbers and types of microorganisms found in a variety of foods.

59. Quantitative Examination of Bacteria in Raw and Pasteurized Milk

Because milk is perishable and potentially a source of disease, its bacterial content is of prime importance. **Pasteurization** is a heat treatment that destroys all pathogenic organisms in milk without completely sterilizing it. Coliforms are always present in raw milk but their presence in pasteurized milk means that the milk has been contaminated after pasteurization.

Two methods, both with advantages and disadvantages, can be used to determine the number of organisms in milk. The agar-plate method, being more sensitive, gives more accurate results for milk containing few bacteria, and the direct microscopic count is more applicable to milk containing many bacteria. Both methods give evidence of the conditions under which milk was collected, handled, and stored, information of considerable significance to the question of sanitation.

Direct plating on desoxycholate agar is both selective and differential for coliforms in milk. This medium inhibits the growth of most bacteria other than the coliforms and related types. The lactose-fermenting coliform bacteria appear as red colonies that can be readily distinguished and counted. Non-lactose fermenters appear as whitish colonies.

This exercise employs the quantitative-plating technique (discussed in Exercise 17) to determine the relative numbers of coliforms and other bacteria in raw and pasteurized milk. The effectiveness of pasteurization is indicated by the reduction of total bacterial numbers as well as by the destruction of all coliforms.

PROCEDURE

1. Plate dilutions of 10^{-2}, 10^{-3}, and 10^{-4} of the sample of raw milk provided on standard-plate-count agar and 10^{-1} and 10^{-2} on desoxycholate agar. *Note*: After the desoxycholate agar has hardened, pour a thin layer of clear agar over the surface to prevent the development of surface colonies that may give atypical reactions.
2. Pasteurize the same milk by heating in a water bath to 61.7°C (143°F) for 30 minutes. Be sure that timing is exact, that the proper temperature is maintained, and that the water level is well above the level of the milk in the container.
3. Plate out the pasteurized milk, using dilutions of 10^{-2} and 10^{-3} on standard plate-count agar, and 1 ml and 10^{-1} dilutions on desoxycholate agar.
4. Incubate all plates at 37°C for 24 hours and make counts of the raw and pasteurized milk on each type of agar.
5. Record your results on the report sheet.

QUESTIONS

1. Why is the square-centimeter grid etched into the glass counting plate on the colony counter? How many centimeters are there in the surface area of a petri plate whose diameter is 100 mm?
2. What pathogenic microorganisms can be found in milk?
3. What microorganisms served as the reference point in determining the times and temperatures for pasteurization?

Quantitative Examination of Bacteria in Raw and Pasteurized Milk

Name _____

Desk No. _____

Sample	Dilution counted	Number of colonies	Bacteria per ml milk
Raw milk desoxycholate agar			
Raw milk plate count agar			
Pasteurized milk, desoxycholate agar			
Pasteurized milk, plate count agar			

60. *Fermented Foods*

SAUERKRAUT

Fermentation is one of the oldest means of food preservation. In making sauerkraut, which is fermented cabbage, acid is produced by lactic acid bacteria, resulting in the inhibition of bacteria that cause spoilage.

Sauerkraut is prepared by salting shredded cabbage. Alternate layers of cabbage and salt are packed into containers and allowed to ferment. About 3 pounds of salt are used for each 100 pounds of cabbage. The salt draws the juice from the cabbage tissue and a brine results in which the acid-forming bacteria thrive. These bacteria ferment sugars to organic acids, mainly lactic acid, which act as preserving agents and prevent the action of putrefactive bacteria.

PROCEDURE

Your instructor will prepare sauerkraut as follows.

1. Halve and core a head of cabbage.
2. Shred the cabbage and alternate layers of cabbage and salt in a container. The amount of salt is 3% by weight.
3. Compress the mixture until a layer of juice covers the cabbage. Cover with a board and weigh it down. Then cover the top of the container with clean cheesecloth.
4. Incubate at 30°C.

You begin work here:

5. Examine the kraut after 2 days, after 1 week, and after 2 weeks of incubation:
6. Taste and smell the kraut. Record descriptions of the flavor and aroma. Prepare a stained preparation of the juice, using a methylene blue stain. Examine and draw the characteristic flora.

Although the development of the fermentative flora might appear to be a chance occurrence, a combination of several environmental factors is at work, including the presence of a fermentable sugar, salt, anaerobiosis, acid pH, and temperature, all of which combine to select the essential organisms for sauerkraut production. The fermentative flora changes as the pH decreases and at varying stages must include *Leuconostoc mesenteroides* and *Lactobacillus plantarum* for proper development of flavor and smell.

BEVERAGES

Fermentation preserves some food and improves the flavor and nutritional value of some others. Most fermented foods require the growth and activity of one or more species of lactic acid bacteria or yeast, or a combination of both. The type of fermentation a food undergoes is determined primarily by its carbohydrate content and pH. Carbohydrate-containing, highly acidic, and poorly buffered foods are conducive to the alcoholic fermentation of yeasts; carbohydrate-containing, low acid, and well-buffered foods commonly undergo lactic acid fermentation. In this exercise you will examine the microflora of some fermented foods.

PROCEDURE

Cultured Buttermilk

Cultured buttermilk is skimmed, or partly skimmed, pasteurized milk that has undergone fermentation by lactic-acid bacteria to yield lactic acid and volatile substances, accompanied by curd production and the development of a pleasing aroma and flavor. It is prepared by souring skim milk with a mixed culture of lactic-acid bacteria, the "starter." When the culture develops about 0.8%–0.9% lactic acid, the fermentation is stopped by cooling, and the curd is broken up by agitation.

Commercial starters usually contain *Lactococcus lactis, Leuconostoc dextranicum*, and *Leuconostoc citrovorum*. The members of the *Leuconostoc* genus attack the citric acid in milk to form diacetyl, which imparts the characteristic aroma to the fermented milk.

1. Inoculate a large tube or flask of pasteurized skim milk with a 1% volume of the starter culture furnished. Distribute the inoculum by rolling the tube rapidly between your hands or agitating the flask.
2. Incubate at room temperature (in your desk) until the next laboratory period (preferably 15–18 hours at 21°C).
3. Agitate the soured milk to break up the curd. Note the aroma and flavor of the cultured buttermilk.

Starter cultures are available from various companies in powdered form. These must be carried through several transfers in milk before a properly balanced product is obtained.

Alcoholic Fermentation of Fruit Juices

A moist acidic medium containing fermentable carbohydrate is likely to undergo alcoholic fermentation by yeasts. Most fruit juices fit this description, and at temperatures of 20–30°C support the growth of yeasts. The fermentation instigated by the common brewing and wine yeasts (*Saccharomyces*) proceeds in accordance with the general equation:

$$C_6H_{12}O_6 \longrightarrow 2\ C_2H_5OH + 2\ CO_2 \uparrow$$
hexose sugar   ethyl alcohol

The term "wine" is most often applied to the product resulting from the alcoholic fermentation of grape juice, but other fermented fruit juices, and even fermented plant juices, are also called wines. There are berry wines, pear, citrus fruit, and dandelion wines, and a host of others. This experiment illustrates the fermentation of grape juice and apple juice to simple wines.

1. Inoculate a tube of the grape-juice medium and a tube of apple-juice medium provided with 0.1 ml of a broth culture of *Saccharomyces cerevisiae*, variety *ellipsoideus*.
2. Incubate at 25°C until the next laboratory period.
3. Note the aroma and flavor of the product.

CHEESE

Various microorganisms are responsible for the particular aromas, flavors, and other characteristics of the various types of cheeses. Each cheese has its characteristic flora. Although the identification of organisms is impractical here, the object of this experiment is to determine the general nature of the characteristic flora of several types of cheese by plating out samples of emulsified cheese. After being weighed, cheese can be prepared for plating by blending with sterile water in a sterile blender.

PROCEDURE

1. Plate a sample of the emulsified cheese provided on bromcresol-purple lactose agar in dilutions of 10^{-2}, 10^{-3}, and 10^{-4}.
2. Incubate the plates until the next laboratory period (2 days) at 30°C.
3. Count the colonies and report in terms of organisms per gram of cheese.
4. Store the plates in your desk for 1 week.
5. Observe the types of colonies from the various cheeses and prepare gram stains from some of them.
6. Observe the textures, aromas, and flavors of the cheeses.
7. Compile your observations in the table on the report sheet.

PICKLES

Pickle products complement many meals or snacks, contribute some nutritive value, are low in fat and, with the exception of the sweet type, are low in calories. There are four general classes of pickle products, classified on the basis of method of preparation and ingredients: brined pickles, fresh-pack, fruit pickles, and relishes. Brined pickles entail a fermentation step in their manufacture.

PROCEDURE

First Laboratory Period

1. Wash the cucumbers gently but thoroughly and remove any blossoms. When the cucumbers are dry, place 1 tablespoon of pickling spices and 3 tablespoons of dill in a 2.2 liter pickle jar; add the cucumbers. If the cucumbers are very large, slice them in half length wise. Place a tablespoon of dill and a garlic clove on top of the cucumbers.

2. Mix 225 ml vinegar and 40 g NaCl in 200-ml of water. With a 1 ml pipet, suspend the starter culture (*Lactobacillus plantarum*) in some vinegar-brine. Pour the inoculum and brine over the cucumbers. Fit the fiberglass screening over the jar contents so that it wedges the cucumbers down in the liquid. Add enough tap water to fill the jar. Cover the jar with the plastic cover, fasten it with the rubber band and incubate at room temperature.

Subsequent Laboratory Periods

3. After 3 to 5 days, scum will appear. Without disturbing the pickles, remove the scum and, add water whenever necessary. In about 3 weeks the cucumbers will be ready to sample.

4. Make gram stains (Exercise 13) at different times during the fermentation to observe the microbes that appear. *Note*: If the pH of the brine has not decreased to pH 4.5 or lower, the pickles may not be safe to eat.

QUESTIONS

1. Many organic substances, for example, raw milk at room temperature or sugar in fermenting fruits, naturally undergo a series of transformations brought about by sequences of microorganisms. Within limits, the appearance of these microorganisms is predictable. What are the chemical changes in the two examples given, and what microorganisms cause them?

2. What is the microbial flora of yogurt? Of Bulgarian buttermilk? Of kumiss?

3. What fermentation end product gives pickles and sauerkraut their characteristic flavors?

4. What distinguishes commercial "light wine" from "fortified wine"?

5. In beer manufacture, why is the grain exposed to the action of germinating barley?

6. What is meant by an acid curd cheese? A rennet curd cheese? A hard cheese? A soft cheese? Name examples of each.

7. Originally, cheese manufacture depended upon variations in environmental conditions to select out the microbial flora for producing a certain cheese. Modern cheese manufacture employs starter cultures and controlled environmental conditions. What are some of these environmental conditions?

REPORT 60

Fermented Foods

Name _____

Desk No. _____

Type of cheese	Characteristics of cheese	Organisms per gram of cheese	Gram stain and morphology of flora
Cheddar			
Blue			
Cottage			
Roquefort			
Camembert			

Discuss the taste and flavor of the foods you prepared in class. How do they compare with those you can buy?

61. *Detecting* Salmonella *on Poultry*

Fluorescent-antibody techniques (FA) are widely used in diagnostic microbiology because they are rapid, sensitive, simple (once the technique has been mastered), and relatively inexpensive per test (after the basic equipment has been obtained). Fluorescent-antibody technique can be used without isolating pure cultures, thus shortening or eliminating cultural steps. For example, fluorescent antibodies that are specific for a pathogen can be used to determine whether the pathogen is present in the mixed flora in sputum or throat washings.

Two staining techniques are most commonly used: the direct, and the indirect. In the direct method a fluorescent dye, usually fluorescein isothiocyanate (FITC), is chemically bonded to the antibody for the pathogen of interest. The antibody is reacted with the antigen and the antibody–antigen complex is observed in the fluorescence or epifluorescence microscope (Figure 61-1).

In the *indirect method* a second step is needed for fluorescent observation. The antibody, which is not tagged with FITC in this method, is first mixed with the antigens. After suitable incubation, the antigen and reacted antibody, is centrifuged and the excess antibody is decanted. The antibody–antigen complex is resuspended in fresh buffer to dilute any entrained unreacted antibody and then centrifuged again. This sediment is resuspended and then reacted with a second antibody that *has* been tagged with FITC and that is specific for the first antibody (Figure 61-2). For example, the first antibody might be made in a rabbit, and the second might be an anti-rabbit antibody that is made in a goat.

You might ask why anyone would do the extra steps of the indirect procedure when direct methods are possible. The answer is that it is relatively easy to make antibodies and more complicated, time consuming, and expensive to bond FITC to antibodies and purify them. If large amounts of diagnostic antibody are being made, it is economical to bond them with FITC, but otherwise the relatively low cost and availability of FITC-tagged antibodies dictate using the indirect procedure.

Fluorescent antibodies are available for diagnosing *Corynebacterium diphtheriae*, enteropathogenic *E. coli* (EEC), *Streptococcus* spp., *Salmonella* spp., *Staphylococcus aureus*, and other pathogens.

In this exercise you will become detectives who are looking for the human pathogen *Salmonella*, which is commonly found on the surfaces of poultry, meat, and in contaminated water. Polyvalent or panvalent antisera against selected somatic and flagella antigens of *Salmonella* that are tagged with FITC will be used to detect them. With slight modifications (e.g., see DIFCO Manual, 10 ed.), the same procedure can be used if other antibodies and antigens are available in the teaching laboratory for a direct FA exercise.

PROCEDURE:

1. Your instructor will give you a sterile swab and saline to take home for samples or you will sample a piece of fresh chicken or turkey skin that you will bring to the lab. Aseptically remove the covering of the swab, moisten it with saline and twirl the swab on the surface of the poultry. Use the swab to inoculate selenite cystine broth (Figure

Fluorescent antibody   Antigen   Fluorescent antigen–antibody complex

Figure 61-1   Direct fluorescent antibody technique.

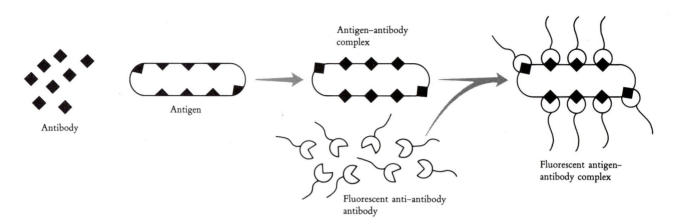

Figure 61-2 Indirect fluorescent antibody technique.

61-3). If you or your instructor choose to sample other foods, two references are available for the appropriate methodology: *Compendium of Methods for the Microbiological Examination of Foods*, 1976, APHA: Washington, DC and *Official Methods of Analysis of the Association of Official Analytical Chemists*, 17th ed., 1989, AOAC: Washington, DC.

2. Incubate the selenite cystine broth for 24 hours ± 2 hours at 35°C.

3. Dilute or concentrate the specimen to approximate a McFarland Barium Sulfate Standard No. 1.

4. Use a 2 mm bacteriological loop to transfer the sample to a clean microscope slide with an etched edge and smear the specimen on the slide, etched side up.

5. Let the slide air dry. Do *not* heat-fix the slide! *Caution:* Do not get the fixative in Step 6 on your fingers!

6. Fix the slide by immersion for 1 minute in FA Kirkpatrick fixative in a Coplin jar (or equivalent).

7. Remove the slide and allow it to drain dry.

8. Apply several drops of the antiserum (FITC conjugate) over the fixed smear.

9. Carefully place the slide in a moist chamber and incubate it at room temperature for 30 minutes. (A moist chamber can be made in a petri dish by placing a paper towel on the bottom and saturating it with water. Bend a glass rod into a V and support the slide on it in the bottom of the dish.)

10. After incubation, drain the excess conjugate and rinse it with FA buffer.

11. Place the slide in a Coplin jar containing FA buffer for 10 minutes. Change the buffer twice during this period.

12. Dip the slide for 30 seconds into a Coplin jar with distilled or deionized water to remove residual salt.

13. Put the edge of the slide onto a paper towel to drain of excess water and then add a small drop of FA mounting fluid.

14. Carefully bring a cover glass to the edge of the drop of the mounting medium. After they have made contact, slowly lower the cover glass to avoid making bubbles between the slide and cover glass.

15. Place the slide in the fluorescent microscope and follow the directions in Exercise 5 with whatever modifications are given to you by your instructor. Use the exciter/barrier filter combination appropriate for FITC.

16. Illustrate your observations on the report sheet.

QUESTIONS

1. Why is it important to detect the presence of *Salmonella* spp. in food?

2. What can be done to retard the growth of *Salmonella* in food?

3. What conditions promote the growth of *Salmonella* in food?

4. Can you name three diseases other than food poisoning that are caused by the growth of microorganisms in contaminated food?

Incubate 24 hrs 35°C

Check OD dilute or
concentrate = McFarland #1

Poultry

Selenite cystine broth

Fixative 1 min

Air dry

Paper towel

Add antiserum

Incubate in moist chamber

Paper towel

Rinse with buffer

30 sec distilled water

Buffer 10 min
2 changes

Paper towel

Cover glass

Mounting medium

Fluorescence
microscope

Figure 61-3 Flow chart of direct fluorescent antibody procedure.

Detecting *Salmonella* on Poultry

Name _____

Desk No. _____

Draw the cells you observed in the fluorescent microscope.

Ask all the other students in your laboratory whether they found *Salmonella* on their poultry.

Class Results

Total Samples   _____
Total Positive   _____
% Positive   _____

F O U R T E E N

Protist Diversity

Advances in fine-structure analysis and in cellular and molecular approaches to diversity among the eukaryotic microorganisms (also known as **protists** or **protoctists**) have blurred the major distinctions among them that microbiologists once made and have changed the methods by which we attempt to classify them. From our late twentieth-century perch, the old problems of separating the protists into the three traditional groups – Algae, Protozoa, and Fungi – seem amusing, but they were practical. The classical methodology stated that if the organism had chloroplasts it was an **alga**, if the organism moved and ingested food it was a **protozoan**, and if the organism was nonphotosynthetic, had a cellulose cell wall, and absorbed nutrients, it was a **fungus**. Those general lines were never very definitive, and some organisms could fit easily into two of the traditional groups. For example, many of the pigmented flagellate groups have representatives that are pigmented and others that are not, some species that are motile and close relatives that are not. Today we find it convenient to use terms such as protozoa (spelled with a lowercase *p*) to refer to the groups that were traditionally included in the former phylum. Although the proper classification of protists may seem complex to a neophyte scientist, it should not serve as a deterrence to observing and enjoying these organisms.

Among the many interesting papers and books on the taxonomy of the protists are the following:

Cavalier-Smith, T. 1987. "The Origin of Eukaryotic and Archaebacterial Cells." In *Endocytobiology III*, J. J. Lee and J. F. Fredrick, Eds. *Ann. N.Y. Acad. Sci. 503*:17–55.

Corliss, J. O. 1983. "A Puddle of Protists." *The Sciences 23*:34–39.

———— 1983. "Consequences of Creating New Kingdoms of Organisms." *Bioscience 33*:314–318.

Lee, J. J., Hutner, S. H., and Bovee, E. C. 1985. *An Illustrated Guide to the Protozoa.* Society of Protozoologists, Lawrence, Kansas.

Margulis, L. 1981. *Symbiosis in Cell Evolution.* W. H. Freeman and Company, San Francisco.

————, Corliss, J. O., Melkonian, M., and Chapman, D. J. 1989. *Handbook of Protoctista.* Jones and Bartlett, Boston.

————, and Schwartz, K. V. 1988. *The Five Kingdoms*, 2nd ed. W. H. Freeman and Company, New York.

BEGINNERS' GUIDE TO MEMBERS OF PROTIST GROUPS

At the present time, microbiologists recognize approximately 50 **phyla** of protists. Discrimination among the phyla is based largely on structural differences, though other criteria can also be important. Many distinguishing features can be seen only in the TEM or SEM, but most phyla can be recognized on the basis of light microscope observations.

One of the most useful ways of sorting out motile phyla is to observe their methods of moving. We can distinguish protists that use amoeboid movement (phyla Rhizopoda, Actinopoda, Foraminifera, Caryoblastea, Acrasiomycota, Labyrinthulomycota, Myxomycota, and Chytridiomycota) from those that move largely by means of flagella (phyla

Dinoflagellata, Praisinophyta, Euglenophyta, Cryptophyta, Parabasala, Kinetoplastida, Choanoflagellida, Rhaphidophyta, Haptophyta, Chrysophyta (some), Chlorophyta (some), Xanthophyta, Eustigmatophyta), those that glide (many pennate Bacillariophyta), those that move by means of cilia (phylum Ciliophora), and those that do not have motile stages during most of their life cycles (phyla Oomycota, Apicomplexa, Hyphochytridomycota, Plasmodiophoromycota, Cnidosporidia, Rhodophyta, and Phaeophyta).

Neophyte protistologists can hardly be expected to recognize members of even this abbreviated list of phyla, but they can appreciate some of the criteria that are used to sort types and classify them.

AMOEBOID PROTISTS

1. Are the pseudopodia very fine hairs or webs?
    (a) Yes. The organism belongs to the
        i. Foraminifera (Figure XIV-1*E*)
        ii. Actinopoda (Figure XIV-1*H*, *I*)
    (b) No. The organism belongs to the
        i. Rhizopoda (Figure XIV-1*A*–*D*, *G*, *J*–*L*, *N*, *O*)
        ii. Karyoblastea (Figure XIV-1*B*)
        iii. Acrasiomycota
        iv. Labyrinthulomycota (Figure XIV-1*M*)
        v. Myxomycota (Figure XIV-1*F*)

If we look at the Rhizopoda (they are common in many freshwater and marine habitats), we find that even the naked amoebae have characteristics that help us distinguish them from each other. For example:

2. Is the pseudopodia type
    (a) broad and cylindrical (e.g., *Hartmannella*, Figure XIV-1*O*)
    (b) slender and cylindrical (e.g., *Vexillifeura*, Figure XIV-1*J*)
    (c) filose (e.g., *Nuclearia*, Figure XIV-1*A*)
    (d) fingerlike (digitiforme) (e.g., *Mayorella*, Figure XIV-1*C*)?
3. Is the body
    (a) flattened (e.g., *Vexillifeura*, Figure XIV-1*J*)
    (b) cylindrical (e.g., *Pelomyxa*, Figure XIV-1*B*)
    (c) polypodal (e.g., *Amoeba*, Figure XIV-1*D* and *G*)
    (d) monopodal (e.g., *Hartmannella*, Figure XIV-1*O*)?
4. Do the pseudopods flow
    (a) continuously (e.g., *Hartmannella*, Figure XIV-1*O*)
    (b) by eruptive locomotion (*Vahlkampfia*, Figure XIV-1*N*)?
5. Is there a recognizable posterior end or uroid?
    (a) Yes (e.g., *Chaos*, *Amoeba*, *Pelomyxa*, Figure XIV-1*B*–*D*)
    (b) No (e.g., *Nuclearia*, Figure XIV-1*A*)
6. Does the amoeba have a test (shell)?
    (a) Yes (e.g., *Euglypha*, Figure XIV-1*L*)
    (b) No (e.g., *Amoeba*, *Chaos*, *Pelomyxa*, Figure XIV-1*B*–*D*)

FLAGELLATED PROTISTS

Flagellated protists are distinguished by many criteria such as the number and type of flagella and the locations of their insertion on the body surface, structures associated with the flagellum, the kinds of waves propagated by the flagella and the direction of movement of the cell, types of chloroplasts (if any), types of storage products, nuclear structure and division, and distinctive surface structures and scales. Some of the features,

A

B

C

D

E

F

G

H

I

J

K

L

M

N

O

Figure XIV-1   Amoeboid protists.
A: Nuclearia.
B: Pelomyxa (Karyoblastea).
C: Mayorella.
D: Amoeba proteus.

E: Foraminifera.
F: Myxomycota.
G: Amoeba proteus.
H: Heliozoan (actinopod).

I: Radiolarian (actinopod).
J: Vexillifeura.
K: Echinamoeba.
L: Euglypha.

M: Labyrinthula.
N: Vahlkampfia.
O: Hartmannella.

such as whether the flagella have fine hairs or scales, are observable only in the electron microscope, but the neophyte can recognize representatives of most common phyla. It must be mentioned for completeness that many of the flagellated phyla have some representatives that have chloroplasts and others that do not; and some are representatives of those protists which are motile only a small part of their life cycles.

The questions below introduce the student to many of the common flagellates (Figures XIV-2 and XIV-3).

1. Does the organism lack chloroplasts? If so, it could belong to one of the following groups:
   (a) Parabasalia. A diverse group of largely parasitic and symbiotic apically multiflagellated protists with characteristic golgi (parabasal bodies) and microtubular rods (axostyles) (e.g., *Tritrichomonas* and *Trichonympha*, Figure XIV-2B,F).
   (b) Kinetoplasta. Apically or subapically uni- or biflagellated, largely parasitic protists, each with a characteristic enlarged, derived mitochondrion (kinetoplast) (e.g., *Trypanosoma*, Figure XIV-2C).
   (c) Choanoflagellata. Free-living attached uniflagellated protists. Flagellum attached apically and surrounded by a funnel-shaped collar formed from tentacles (microvilli). Often surrounded by a theca or a basketlike lorical made of siliceous ribs (costae) (e.g., *Codosiga*, Figure XIV-2K).

2. Does the organism have chloroplasts or is it a relative of organisms that have chloroplasts?
   (a) Is the organism green (chlorophyll *a* and *b* and various carotenoids)?
       i. Euglenophyta. Flagella arise from within an anterior invagination of the cell, the reservoir (tubular canal and pyriform chamber). The basic number is two heterodynamic flagella per cell, but in some species only one emerges from the canal. A few parasitic species have more than two flagella. Most species have an orange-red eyespot. The pellicle consists of interlocking proteinaceous strips that can slide relative to each other in a characteristic rapid change in shape known as *euglenoid movement*. Euglenoids vary in their modes of locomotion. Some swim rapidly while rotating in a helix, others glide with an outstretched anteriorly directed flagellum (the second flagellum trails, if present), and others creep (e.g., *Euglena* and *Peranema*, Figure XIV-3BB, CC).
       ii. Chlorophyta. Two, sometimes four, smooth, anteriorly directed, equally long, homodynamic flagella attached anteriorly, often through a raised papilla. Many have a single cup-shaped chloroplast with a single basal pyrenoid and a red eye spot. Many subgroups have nonmotile forms. Colonial forms found in some subgroups (e.g., *Chlamydomonas, Gonium, Volvox, Chlorococcum, Chlorogloea, Ulva, Acetabularia, Micrasterias, Spirogyra* and *Sorastrum*, Figures XIV-2L, N–U, and XIV-3KK).
       iii. Prasinophyta. One, two, four, or eight scale-covered flagella that may arise from the bottom of a deep pit. Usually one cup-shaped or lobed chloroplast per cell with one or more large pyrenoids. Bright red eyespot in most species. Cell body covered by one or more layers of scales (e.g., *Pyramimonas*, Figure XIV-3FF).
   (b) Is the organism brown, olive, or reddish-brown (chlorophylls *a* and *c* and various carotenoids)?
       i. Dinoflagellata. Two heterodynamic flagella, one of which is ribbon-shaped with a paraxial rod and a single row of fine hairs (seen in the TEM) and the other with two rows of stiffer hairs (seen in the TEM). The ribbon flagellum is usually wrapped transversely around the body, and the other flagellum trails. The organisms characteristically spin as they move through the water. Their nucleus is unique among eukaryotes because the chromosomes are always condensed and their DNA is not complexed with protein. In a phase-contrast microscope, the chromosomes look like beads. The cells are often flattened, or of complex symmetry, and have transverse and ventral grooves in which the flagella lie. Many groups have very thick cellulosic plates (e.g., *Gymnodinium, Gyrodinium, Amphidinium, Ceratium,* and *Peridinium*, Figure XIV-2A, G–J)
       ii. Cryptophyta. One or two boat-shaped or bilobed chloroplasts that are olive-green, brownish, or sometimes red or bluish. Flagella arise on the dorsal face of an apical, nearly apical, or ventral invagination (gullet). There are two unequal flagella, the longer of which carries a bilateral array of stiff hairs (seen only in TEM or SEM) and the shorter only one row (seen only in TEM or SEM). The periplast (outer envelope) consists of internal plates with upturned edges. Trichocysts are present under each corner of the periplast plates and in the gullet (e.g., *Cryptomonas*, Figure XIV-2M).

A

B

C

D

E

F

G

H

I

J

K

L

M

N

O

P

Q

R

S

T

U

Figure XIV-2   Genera of flagellated protists.

A: *Gymnodinium.*
B: *Tritrichomonas.*
C: *Trypanosoma.*
D: *Tetramitis.*
E: *Hexamita.*
F: *Trichonympha.*

G: *Amphidinium.*
H: *Gyrodinium.*
I: *Peridinium.*
J: *Ceratium.*
K: *Codosiga.*

L: *Volvox.*
M: *Cryptomonas.*
N: *Chlamydomonas.*
O: *Gonium.*
P: *Micrasterias.*

Q: *Chlorogloea.*
R: *Chlorococcum.*
S: *Spirogyra.*
T: *Ulva.*
U: *Acetabularia.*

353

Figure XIV-3   More genera of flagellated protists.
AA: *Chrysochromulina*.   DD: *Synura*.   GG: *Emiliania*.   JJ: *Chrysodendron*.
BB: *Euglena*.   EE: *Ochramonas*.   HH: *Dinobryon*.   KK: *Sorastrum*.
CC: *Peranema*.   FF: *Pyramimonas*.   II: *Chrysonema*.

iii. Chrysophyta. Cells usually have two unequally long heterodynamic flagella. The longer is
directed anteriorly and the shorter directed laterally when emergent. In the electron microscope,
the longer flagellum bears bilateral tubular mastigonemes and the shorter one is smooth. An
orange-red eyespot is found in most species. Pigmented forms usually have two yellow-brown or
yellow-green chloroplasts. Most Chrysophytes are naked though some genera are covered with

siliceous scales. Most species are free-swimming, but colonial, rhizopodial, palmelloid, coccoid, and filamentous forms also belong to the group (e.g., *Ochromonas, Chrysodendron, Dinobryon, Chrysonema,* and *Synura,* Figure XIV-3DD, EE, HH–JJ).

iv. Prymnesiophyta. These organisms bear a unique, thin filiform organelle, the *haptonema,* which is inserted between two posteriorly directed smooth flagella. Usually two golden-brown saucer-shaped parietal chloroplasts. Most species do not have an eye spot. Most species are motile flagellates, but amoeboid, palmelloid, colonial, and filamentous forms also belong to the group or are part of the life cycle. The cell covering consists of one or more layers of unmineralized or mineralized (calcite or aragonite) scales (e.g., *Chrysochromulina* and *Emiliania,* Figure XIV-3AA, GG).

CILIATED PROTISTS

All ciliated protists belong to a single phylum, all are dikaryotic (have a macronucleus and a micronucleus; (see Color Plate 7). Many of the features used in the classification of ciliates are best seen after the organisms are stained with protargol (Lee et al. 1985). Ciliates are distinguished from one another by the structure of their oral and somatic regions. The basic structure of their cortex is the **kinetid,** which consists of one, two, or many **kinetosomes** and their associated cilia, fibrils, and ribbons, all of which can be seen with good silver stains. Some common orders are listed to give you an idea of the diversity in the group (Figure XIV-4):

1. Protostomatida. No mouth, ingestive area apical or a long naked ventral stripe (e.g., *Tracheloraphis,* Figure XIV-4A).
2. Loxodida. Cilia mainly on the right surface of flattened body and a tubular oral cavity with simple cilia (e.g., *Remanella,* Figure XIV-4B).
3. Armophorida. Body twisted to left with oral region spiralled. In protargol, one can see 3–5 perioral kinetics along the edge of the oral region (e.g., *Caeomorpha,* Figure XIV-4G).
4. Heterotrichida. Body has two kinetosomes (dikinetids) with cilia arising from forward kinetosome. Left serial oral polykinetids used to locomote and feed. Some have a lorica (e.g., *Stentor,* Figure XIV-4N and Color Plate 8).
5. Choreotrichida. Body generally conical or bell-shaped with little body ciliature. Oral ciliature around broad end arises from polykinetids. Usually with a lorica (e.g., *Cyttarocylis,* Figure XIV-4L).
6. Stichotrichida. Body cilia as files or scattered groups of *cirri* (polykinetids). Left oral cilia as a "collar" and "lapel" of polykinetids (e.g., *Urostyla,* Figure XIV-4E).
7. Haptorida. Cytopharynx is not inverted. An oral dome is supported by two sets of microtubules that arise from oral dikinetids (e.g., *Didinium,* Figure XIV-4K).
8. Suctoria. Sessile adult forms (trophonts) capture and ingest food with one or more tentacles. Larval, migratory stage, is ciliated (e.g., *Tokophrya,* Figure XIV-4M).
9. Peniculida. Have an oral cavity that is often very deep, with three left oral kinetids with the long axis parallel to the long axis of the oral cavity (e.g., *Paramecium,* Figure XIV-4J).
10. Hypotrichia. Body is ovoid and dorsoventrally flattened. Ciliature on dorsum; cilia on the ventral side grouped into tufts (cirri) (e.g., *Euplotes,* Figure XIV-4H).
11. Hymenostomatia. Oral cavity with a distinctive ciliature consisting of a large right oral membrane and three left oral membranelles (e.g., *Tetrahymena,* Figure XIV-4C).
12. Peritrichia. Prominent oral area bordered by a kinetid file and an oral kinetid in the oral cavity (infundibulum). Ciliates in this group may be attached by stalks, live in a lorica, or live symbiotically in or on other organisms. Some are colonial (e.g., *Vorticella,* Figure XIV-4O).

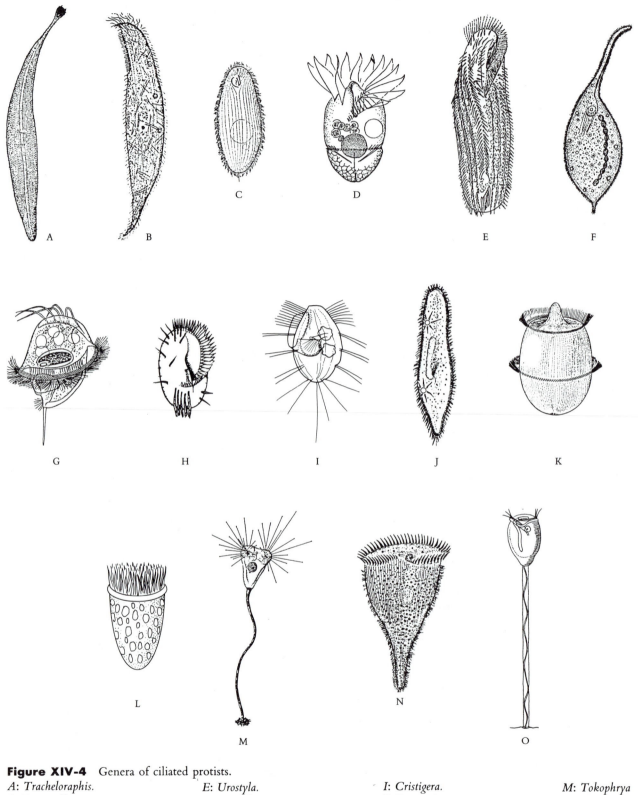

Figure XIV-4   Genera of ciliated protists.

A: *Tracheloraphis.*
B: *Remanella.*
C: *Tetrahymena.*
D: *Strombidium.*

E: *Urostyla.*
F: *Paradileptus.*
G: *Caeomorpha.*
H: *Euplotes.*

I: *Cristigera.*
J: *Paramecium.*
K: *Didinium.*
L: *Cyttarocylis.*

M: *Tokophrya*
N: *Stentor.*
O: *Vorticella.*

Figure XIV-5 Genera of diatoms. *A*: *Asteromphalus*. *B*: *Skeletonema*. *C*: *Opehora*. *D*: *Fragilaria*. *E*: *Navicula*. *F*: *Cocconeis*. *G*: *Nitzschia*. *H*: *Mastogloea*. *I*: *Diploneis*. *J*: *Amphora*. *K*: *Rhopalodia*.

GLIDING AND NONMOTILE PROTISTS   Diatoms (Bacillariophyta) are the most abundant photosynthetic aquatic protists. They are pigmented brown and have chlorophylls *a* and *c*. Diatoms have siliceous frustules (shells) made of two interlocking parts (valves), which are extremely elaborate. The protoplasm must be removed (Exercise 63) to study the frustule structure, which is used as the basis for classifying and identifying the estimated 1×10^4 living species in the phylum.

Diatoms are divided into two major subgroups by the overall symmetry of the valves (Figure XIV-5):

1. Centric. Is the symmetry of the valve organized in reference to a central point in the valve (centric or radial valves) or in reference to two, three, or more points (biangular, triangular, or polygonal valves)? (See Color Plates 9 and 10)
    (a) Coscinodiscales. Organisms with flat or convex disc-shaped valves without prominent processes or bands (intercalary), which increase thickness. The diameter of the frustules is greater than thickness (e.g., *Actinocyclus*, Color Plate 11).
    (b) Rhizosolenales. Frustules long and cylindrical with many bands so that the diameter of the frustule is much smaller than its thickness (girdle view) (e.g., *Melosira*, Color Plate 12).
    (c) Biddulphiales. Frustules, which have bi- or multipolar valves with angles, provided with spines, elevations, ocelli, or other projections (e.g., *Biddulphia*, Color Plate 13).
2. Pennate. Is the structure of the valve arranged in reference to a central line? [Frustules with a slit (raphe) usually glide] (Figure XIV-5H).
    (a) Araphidales. Organisms without raphe (e.g., *Lichmophora*, Color Plate 14).
    (b) Raphidioidales. Organisms with a rudimentary raphe at the ends of the valves.
    (c) Monoraphidales. Organisms with a raphe on only one of the two valves and a pseudoraphe (hyaline area) on the other valve (e.g., *Cocconeis*, Figure XIV-5F).
    (d) Biraphidales. Organisms with a raphe on both valves (e.g., *Navicula, Amphora,* and *Nitzschia,* Figure XIV-5E, G, J).

62. *Some Interesting Places To Sample Protists*

AQUATIC HABITATS

Protists abound in every aquatic habitat you can imagine. They can be free swimming or attached to substrates. Some are osmotic, others are photosynthetic, and still others are herbivorous, carnivorous, bactivorous, and even myxotrophic. The characteristics of the habitat determine the types of protists found there. For example, enormous numbers of bactivorous flagellates, ciliates, and amoebae and carnivorous ciliates are found in the oxidation ponds and settling tanks of sewage treatment facilities; large numbers of autotrophic and herbivorous protists are found in well-illuminated oligotrophic waters.

PROCEDURES

Plankton in Lakes, Ponds, Salt Marshes, and Shallow Embayments of the Sea

1. (a) Planktonic protists may be concentrated enough in their natural habitat to be collected in a widemouthed jar and studied directly. Keep your samples out of the

direct sun after they are collected because water temperatures over 30°C kill many protists. Keep samples in a styrofoam insulated chest during transport to your laboratory.

(b) Planktonic protists can be concentrated by collecting a sample using a 50 μm plankton net. Most of the smaller protists are lost this way and many others are damaged, but you will still get a very interesting collection to study under your microscope. Place your samples in an insulated chest for transportation back to your laboratory.

OR

(c) Collect a large sample of water (100 ml) and place it in an insulated chest for transportation to your laboratory.

2. If you want to study live specimens, you can concentrate a sample of the collection by gentle centrifugation. Some organisms may be damaged, but most will survive.

3. If you want to study only the diatoms in your sample, you can separate them out by filtering the water sample through a Millipore filter; then wash the diatoms off the filter into a centrifuge tube. Process the sample as directed in Exercise 63. You can accomplish the same goal by concentrating the sample by centrifugation before processing the sample as directed in Exercise 63.

OR

4. You can fix the sample of organisms with Bouin's fixative [20:1; sample: Fixative (v:v)]. This fixative is prepared by making a saturated solution of picric acid in buffered concentrated liquid formaldehyde. Just before using, add glacial acetic acid (5 ml/100 ml fixative) to the fixative. Other common fixatives are Lugol's iodine, 5% glutaraldehyde, and buffered formalin. No one type of fixative is useful for all types of protists.

5. Mix the fixative with the sample by gently stirring or shaking the sample; then pour the sample into a graduated cylinder.

6. After the organisms have settled to the bottom of the cylinder (this may take 10 minutes to 1 hour), decant the excess fixative.

7. Pour the residue into a centrifuge tube.

8. Now you have three options:
   (a) Place a drop on a microscope slide and cover it with a coverglass.

OR

   (b) Stain the organisms with Protargol (Lee et al. 1985).
   (c) Concern yourself only with the diatoms in the sample. Prepare slides following the protocol in Exercise 63.

9. Study your preparations in a microscope with the aid of the text and figures of the "Beginner's Guide to Identifying Members of Protist Groups," earlier in this section.

Protists in the Shallow Benthos (bottoms) of Lakes, Ponds, and Shallow Embayments of the Sea

1. (a) With a forceps carefully hold and break a small leaf of submerged aquatic microphyte or seaweed and transfer it to a large test tube (25 × 150 mm) containing sterile water of equal osmolarity and pH (e.g., pond water or sea water).

OR

   (b) With a forceps remove some detritus, small pieces of rubble, or similar materials from the surface of the lake or pond sediment and transfer it to a widemouthed plastic jar half full of water from the same site.

OR

   (c) With the aid of an angled kitchen cooking flipper (spatula), carefully skim the upper 1 cm of the surface sediment and transfer it to a widemouthed plastic jar half full of water from the same site.

2. Vigorously shake the sample a few times to dislodge attached protists and suspend them in the water.

3. Decant the sample into a centrifuge tube, leaving behind the large particles of leaves, stones, and sediments.

4. (a) The sample can be examined directly.

OR

(b) The sample can be processed for the study of the diatoms (Exercise 63).

OR

(c) The sample can be fixed as indicated in step 4 or the procedure for studying plankton.

5. Study your preparations with the aid of the text and figures of the "Beginners Guide to Identifying Members of Protist Groups," earlier in this section.

Sponge Colonization Samplers

On a campus where vandalism or theft is not a problem, the sampling method pioneered by J. Cairns and his colleagues is a convenient and simple method of collecting protists for study:

1. Take a piece of polyurethane foam and tie it to a fishing line between a float and a sinker. Any number of plastic sponge islands can be tied along the line, and samples can be obtained from any depth.
2. Let the foam sponge incubate in place for several days to a week.
3. When you are ready to study the protists, squeeze out the sponge into a beaker. You have the same options for studying these specimens as you had in the other sampling methods.

QUESTIONS

1. Are there any explanations of the great diversity of protists in aquatic habitats?
2. What is the meaning of the ecological term *community*? Do true communities form in the plankton of aquatic habitats?
3. Make a list of aquatic habitats in which each of the following might be characterized as the dominant protists: bacteriovores; herbivores; and mixed communities of photosynthetic protists, osmotrophs, herbivores, and carnivores.
4. What are the most important factors that affect the structure of the microbial food webs in planktonic communities?
5. What are the most important factors that affect the structure of the microbial assemblages in the shallow benthos of lakes or at the edges of the sea.

PROTOZOA OF COMMON LABORATORY ANIMALS

Almost all animals are hosts for parasitic and commensial protozoa. Most are found in the midgut and hindgut. Common laboratory animals are cockroaches, frogs, and rodents (mouse or rat). Any of these will provide examples of these protists. If termites or rumen contents (see Color Plate 15) are available, they also provide interesting material for study.

PROCEDURE

Cockroach

1. Anesthetize a cockroach with ethyl ether and pin it ventral side up in a dissecting pan.
2. With a sharp dissecting scissor, cut open the abdomen of the roach while at the same time bending the roach with your other hand so that the gut is exposed.
3. Grasp the intestinal tract with a fine forceps and separate it from the rest of the body tissue.
4. Place the intestinal tract on a microscope slide with a drop of tap water.
5. With a pair of dissecting needles, tease apart the midgut of the roach. Gregarines (protozoa, phylum Apicomplexa) are often found there. The most common is *Gregarina blattarum* (Figure 62-1Q) whose sexual stage (gamonts) are large enough (350–500 μm by 185–200 μm) to be seen with the unaided eye. Two other gregarines, *Protomagalhaesia serpentula* and *Plistophora rudoir*, are also found in the cockroach midgut.
6. Using the sides of the dissecting needles to hold and squeeze the hindgut, release the contents onto the slide. Throw away the intestinal tract and place a cover glass over the drop. Alternatively, the drop could be transferred to a centrifuge tube and stained with protargol (Lee et al. 1985) before study.
7. Three types of protozoa—ciliates, flagellates, and amoebae—are found in the hindgut. The ciliates, *Balantidium blattarum* (Figure 62-1C) and *B. praenucleatum* and *Nyctotherus ovalis* (Figures 62-1D, E), are easily identified at low (100×) and medium (400×) magnifications. You need an oil-immersion lens (1000×) to visualize the flagellates *Retortamonas blattarum* (Figure 62-1I),

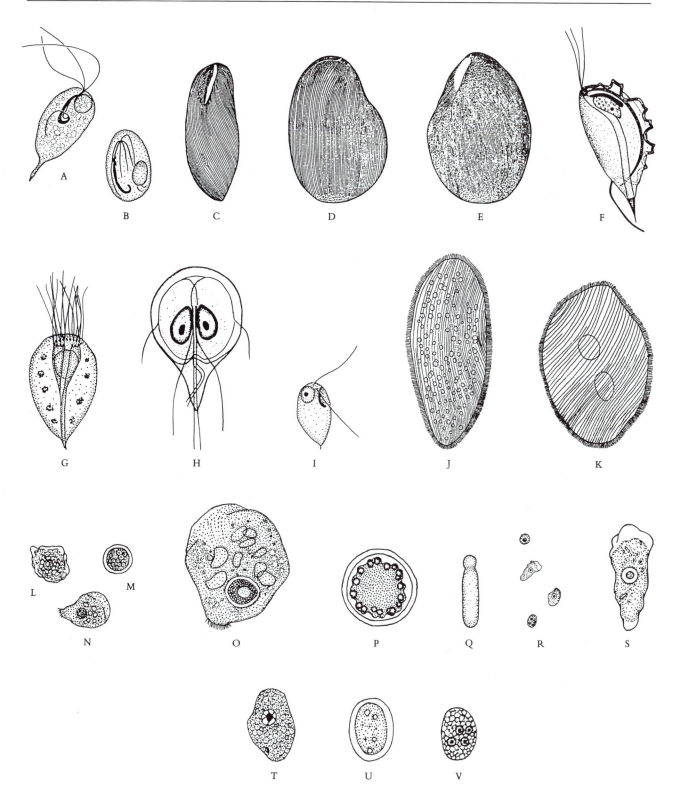

Figure 62-1 Protozoa of common laboratory animals. *A: Chilomastix. B: Chilomastix* cyst. *C: Balantidium. D, E: Nycto-therus* (right and left views; dorsal). *F: Tritrichomonas. G: Lophomonas. H: Giardia. I: Retortamonas. J: Opalina. K: Zelleriella.*
L–P: Endamoeba. Q: Gregarina blattarum. R–V: Endolimax.

Hexamita periplaetae (Figure XIV-2E), *Tetratrichomastix blattaridium, Lophomonas blattarum* (Figure 62-1G), and *L. striatar,* and the tiny amoebae *Endamoeba blattae* (Figure 62-1L–P), *E. thomsoni,* and *Endolimax blattae* (Figure 62-1R–V).

Frog

1. Anesthetize a frog and pin it in a dissecting tray ventral side up.
2. Make an incision in the abdomen, exposing the large intestine.
3. Hold the large intestine with a forceps and cut it open with a scissors. Remove the contents of the large intestine with a Pasteur pipet and transfer a drop to a microscope slide. Cover the drop with a cover glass. As an alternative, you could transfer the contents of the large intestine to a centrifuge and stain them using the protargol procedure (Lee et al. 1985).
4. Two large ciliates, *Balantidium duoden* (Figure 62-1C) and *Nyctotherus cordiformis* (Figure 62-1D,E), and two large flagellates, *Opalina* spp. (Figure 62-1J and Color Plate 16) and *Zelleriella* spp. (Figure 62-1K) are easily detected at low (100×) and intermediate (400×) magnifications. The other flagellates, *Tritrichomonas augusta* (Figure XIV-2B), *Chilomastix caulleryi* (Figure 62-1A), and *Hexamita intestinalis* (Figure XIV-2E), and the amoebae, *Endamoeba ranarum* (Figure 62-1L–P) and *Endolimax ranarum* (Figure 62-1R–V), are too small to be recognized at magnifications of less than 900×.
5. Wild frogs may also have species of *Trypanosoma* (see Figure XIV-2C) (*T. rotatorium* and *tinopinatum*) in their blood. These can be visualized in blood smears stained with Wright's stain.

Rodent

1. Anesthetize a mouse or rat and pin it in a dissecting tray ventral side up.
2. Make a ventral incision and expose the digestive tract. Locate the small intestine, caecum and large intestine.

3. With a forceps hold the small intestine and cut it open with a scissors. With a Pasteur pipet, take a sample of the intestinal contents, and transfer a drop to a microscope slide. Cover the drop with a cover glass. The intestinal flagellate, *Giardia muris* (Figure 62-1H and Color Plate 17) is easily detected at high magnifications (1000×). As an alternative, the small and other intestinal samples could be transferred to centrifuge tubes and stained with protargol (Lee et al., 1985) before study.
4. With a forceps hold the small intestine and caecum with one hand and snip open the caecum with a scissors in the other hand. Remove the contents of the caecum with a Pasteur pipet and place a drop on a microscope slide. Cover with a cover glass. At low (100×) and intermediate (400×) magnifications, you can detect the ciliate *Balantidium caviae* (Figure 62-1C), and at higher magnifications *Tritrichomonas microti* (Figure 62-1F).
5. With a forceps hold the large intestine and cut it open with a scissors. Remove some contents with a Pasteur pipet and place a drop on a microscope slide. Cover with a cover glass. At high magnifications (900×) the flagellates *Tritrichomonas muris* (Figure 62-1F) and *Chilomastix bettencourti* (Figure 62-1A) and the amoebae *Endamoeba muris* (Figures 62-1L–P) and *Endolimax nana* (Figures 62-1R–V) can be detected.
6. The blood of some wild rodents can contain trypanosomes (see Figure XIV-2C) (*Trypanosoma cruzi, T. lewisi, T. duttoni*) and the malarial parasite, *Plasmodium berghei*. These can be visualized by making blood smears and staining them with Wright's stain.

QUESTIONS

1. Why are animal intestines a good habitat for various types of protozoa?
2. Do humans have protozoa in their intestinal tract? Do any of them cause diseases?
3. What do you think are the nutritional sources of the intestinal protozoa?

63. *Preparation of Diatom Frustules for Microscopic Study*

Live diatoms are difficult for a beginning scientist to identify. The brown color of the chloroplasts makes it difficult to discern the morphological features on the surface of the transparent siliceous **frustules**; a frustule is the cell wall of the diatom (see Color Plate 18).

One of the easiest and gentlest methods for cleaning diatoms for microscopical study employs hydrogen peroxide. Other methods use strong mineral acids (e.g., H_2SO_4 and HNO_3).

PROCEDURE

1. *Put on protective plastic gloves and protective glasses or goggles.*
2. Take 5 ml of your sample containing diatoms and place it in a 15 ml clear, conical bottom, centrifuge tube. The sample can be from a plankton tow, epiphytes washed from benthic macrophytes, or scrapings from other submerged substrates.
3. Carefully open and then pipet 5 ml of 10–30% concentrated H_2O_2 into the centrifuge tube containing your sample.
4. Place your centrifuge tube in a rack immersed in a 50°C water bath.
5. Watch the contents of your centrifuge tube. *If they bubble too vigorously, remove the tube from the water bath.* Replace it in the bath when the bubbling dies down. When the sample loses all traces of color and becomes as transparent as the water in the tube, remove it from the bath.
6. Gently centrifuge the tube for 5 minutes (approximately 1000 rpm in table top clinical centrifuges).
7. Carefully remove the supernatant fluid by decanting or by pipet and replace it with distilled water.
8. Centrifuge again and decant. (You may now remove your protective gloves and goggles.)
9. Add 10 ml 95% ethyl alcohol and agitate the sample in a vortex mixer.
10. Centrifuge and decant.
11. Add 100% ethyl alcohol to your sample and then agitate it in a vortex mixer.
12. Centrifuge and decant.
13. *Repeat* Steps 11 and 12.
14. Add 10 ml of toluene to your sample.
15. Centrifuge and decant.
16. Add 10 ml of toluene to your sample again.
17. Centrifuge and decant.
18. With a pipet, take a drop of your sample and place it on a microscope slide.
19. Take a drop of mounting medium and place it next to your sample drop (either Hyrax or Permount are recommended mounting media).
20. Mix the two drops thoroughly and then place a #0 or #1 cover glass at the edge of the combined drop. Slowly lower the cover glass onto the drop, trying not to entrain air bubbles.
21. The slide is now ready to study in a brightfield or phase contrast microscope. Use the text of Exercise 62 and Figure XIV-5 to see how many diatom taxa you can find. The presence of a raphe and its position on the valve (or its absence) is the key character for classifying pennate diatoms.

QUESTIONS

1. What is the purpose of using H_2O_2 in the procedure for cleaning diatom frustules?
2. The step in which 100% ethyl alcohol is added to the sample is considered to be a critical one. Why?
3. Do any of the structures found on a diatom frustule (e.g., raphe) have functions?

64. *Axenic Isolation of Diatoms and Chlorophytes*

Although it is not widely known, it is very easy to isolate **axenic** (pure, not contaminated with other organisms) cultures of diatoms and chlorophytes. These photosynthetic protists are found in the illuminated waters of almost every freshwater and marine habitat. The modifications of the usual pure culture techniques (Exercise 9) that are listed in these procedures will enable you to isolate a great variety of photosynthetic protists from your sample. The method can be further improved by supplementing the medium with vitamins (e.g., biotin, thiamine, or B_{12}) or simple substrates (e.g., acetate, lactate, glycine, or glycerol). A classic paper on this method and its modifications (J.J. Lee et al., 1975, *J. Phycology* 11:14–49) can be consulted for more details.

PROCEDURE

Steps 1–8 may be done by your instructor or laboratory assistant.

1. Collect a sample of water from the pond, lake, or marine habitat from which you intend to isolate your diatoms.
2. Filter the water through #1 filter paper.
3. Divide the water into two fractions: Medium A (unenriched) and Medium B (enriched).
4. Add soil extract (5 ml/100 ml), $NaNO_3$ or NH_4NO_3 (10 mg/100 ml), and $NaH_2PO_4 \cdot H_2O$ (5 mg/100 ml) to 95 ml of the water to make Medium B. Adjust the pH of the water to its original value if it has changed (e.g., coastal sea water should be pH 8.1).
5. Add 15 g/l of a good quality agar (e.g., Noble) to both media. If the agar is not of high quality, agar flakes will make it difficult to see the algal clones on the medium. Boil to dissolve the agar.
6. Autoclave for 15 minutes.
7. Transfer the medium from the autoclave to a water bath at 60°C.
8. Just before pouring the petri plates, aseptically add an antibiotic–antimycotic mixture (1 ml/100 ml; GIBCO Laboratories Catalogue #600-5240) that contains the following: Penicillin 10,000 units ml; Streptomycin 10,000 μg/ml; and Fungizone 25 μg/ml. Mix thoroughly and pour the plates.
9. Take a small sample (<1 cm² surface) of a seaweed or similar aquatic substrate and transfer it to a 100 ml milk dilution bottle containing sterile water of the same osmolarity (e.g., pond water or sea water) as the sample (Figure 64-1). Shake the sample vigorously to dislodge the epiphytes from the seaweed.

   A plankton sample could be used as an inoculum. If it is optically transparent to the eye, it can be used as it is; if it is turbid, it should be diluted so that the concentration of the protists is approximately 1×10^{-2}–1×10^{-3}/ml.
10. Pipet 0.1 ml of the sample onto the surface of the agar plate. Using an alcohol-sterilized glass spreader, cover the surface of the plate with the inoculum. Inoculate two plates each of Medium A and B.
11. Place the inoculated petri dishes into plastic sandwich bags and tape the bags closed.
12. Invert the plates and incubate them over a set of fluorescent lamps. The temperature can be from 18 to 25°C, and the light levels can be from 30 to 50 μ-Ein/m² · /sec (where μ-Ein is micro-einstein).
13. After the plates have incubated for two weeks, remove them from the plastic bags and examine them under a dissecting microscope or the low-power ($5 \times 10 = 50\times$) objective of a compound microscope. You will find that diatoms form a large variety of colony types, which helps the experienced scientist in selecting organisms to study (Figures 64-2*A*–*F*). Some descriptive terms are as follows:
    (a) Compact colonies
        i. Circular
            a. Regular border (Figure 64-2*A*)
            b. Irregular border (Figure 64-2*B*)
                (i) Rhizoid (Figure 64-2*C*)

Old refrigerator shelf
or chicken wire

2 × 4 wood

Fluorescent
fixture

Turntable

Figure 64-1 Flow chart for the isolation of diatoms.

    (ii) Filamentous (Figure 64-2*E*)
    (iii) Lobate (Figure 64-2*D*)
  ii. Rectilinear (Figure 64-2*F*)
(b) Colonies with motile diatoms (center of colony compact and motile diatoms at periphery)
  i. Discontinuous waves of growth at periphery (Figure 64-2*G*)
  ii. Continuous unpunctuated growth with well-separated diatoms (Figure 64-2*H*)

  iii. Groups of diatoms at the edge grouped into
    (a) Stacks (Figure 64-2*I*)
    (b) Packets (Figure 64-2*J*)
    (c) Secondary colonies (Figure 64-2*K*)
14. Choose the well-separated colonies you want to isolate and ring them with a wax pencil on the bottom of the dish.
15. With an alcohol-sterilized small spatula (e.g., Fisher Scientific Co., Catalog 21-401-10), lift the colony and transfer it to liquid

A

B

C

D

E

F

Figure 64-2 Colonial morphology of some pennate diatoms (stereomicroscopic photographs).

G

H

I

J

K

medium of the same type as you used for the initial isolation.

16. Incubate in the light for 7–10 days.

17. Prepare the organisms for study in either the light microscope or the SEM, following the methods described in Exercise 1, 4, or 6.

QUESTIONS

1. Why must your plates be incubated inside plastic bags?

2. Why are the two media selective for the growth of diatoms and chlorophytes?

F I F T E E N

Fungi: Molds and Yeasts

Everyone has observed fungi, cottony or feltlike, growing on foods and other materials. If you examine these molds (Color Plate 19), as they are commonly called, with a simple magnifying lens, you will see a mass of branching intertwining threads, the **mycelium** of the mold. The diameter of a single strand of the mycelium, a **hypha**, is 5–10 times the diameter of the average **eubacterial** cell we have been observing. Much of the mycelium of each mold grows within the growth medium or on its surface and functions to extract nutrients. This is termed the **vegetative mycelium** (Figure XIV-1). Arising from or being produced from portions of the vegetative mycelium are specialized fruiting structures (e.g., **sporangia, asci**, and **basidia**) that produce asexual or sexual spores (Color Plates 20–22). Fungi excrete powerful enzymes that break down food into molecules, which the fungi then can absorb. Nearly all fungi are aerobes. Many fungi are **saprobes**, living on decaying vegetation; others are parasitic or symbiotic (e.g., lichens and mycorrhiza).

There may be as many as 1×10^5 species of fungi, separated into four phyla – Zygomycota, Ascomycota, Basidiomycota, and Deuteromycota – on the basis of their mycelial organization and their sexual and asexual reproductive structures.

The mycelium in the Zygomycota is **aseptate** (lacking cross walls, or **septa**) except between reproductive structures and the rest of the coenocytic mycelium (Figure XV-2). Members of this phylum form their haploid asexual spores, **sporangiospores**, within an aerial sporangium (Figure XV-3). Conjugation occurs when two specialized hyphae, **gametangia**, of opposite mating types grow toward each other until they touch. The gametangia swell, their cytoplasms and nuclei fuse, and a thick-walled **zygosporangium** is formed (Color Plate 21). Meiosis takes place within the zygosporangium, and eventually haploid spores are dispersed. Members of the largest of the subgroups of Zygomycota, Mucorales, are saprobic and live on decaying plants or animals and in dung. Some, such as *Rhizopus stolonifer*, spoil strawberries and sweet potatoes in transit and storage, and some others produce yeastlike forms in liquid media. Some of the more common genera are illustrated in Figures 65-2*A,B*).

Ascomycota are distinguished from other fungi by their sexual reproductive structure, the ascus. An **ascus** is a capsule that is formed by two hyphae of compatible mating types when they conjugate. After the formation of a zygote, the transient diploid nucleus undergoes meiosis, producing 4 new haploid nuclei (there may be a terminal mitotic division or two, resulting in 8 or 16 nuclei), each of which is then surrounded by a protective spore wall (**ascospores**). The mycelium of Ascomycota is **septate** (Figure XV-4).

Asexual reproduction in the Ascomycota is quite varied, involving fission, budding, fragmentation, chlamydospores, or conidia, depending on species and environmental conditions. Fission and budding are methods of asexual reproduction found in yeasts and a few other members of the fungi. Asexual spores produced by budding are called **blastospores**. Most Ascomycota form one of a diverse group of asexual fruiting structures known as **conidia**; the most familiar are borne on hyphal branches known as **conidiophores** (Figure XV-5 and Color Plate 22). There are four major classes of

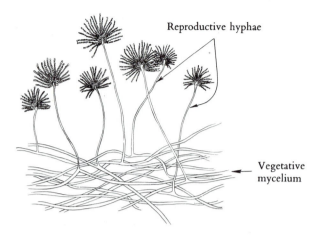

Figure XV-1   Structure of a mold.

Figure XV-2   Coenocytic mycelium.

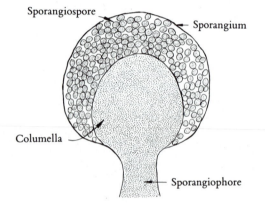

Figure XV-3   The sporangium of a zygomycote.

Figure XV-4   Septate mycelium.

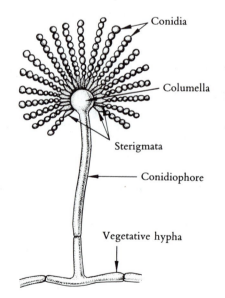

Figure XV-5   The conidium of the ascomycote *Aspergillus*.

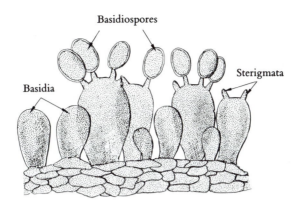

Figure XV-6   Section of a basidiomycote.

Ascomycota: Hemiascomyceta; Euascomyceta; Loculoascomyceta, and Laboulbeniomyceta. The Hemiascomyceta are morphologically simple with short mycelia or none at all; they include the familiar bakers' and brewers' yeasts. The Euascomyceta are the largest and best-known group. The Loculoascomyceta from a typical sexual fruiting body with an apical basket that contains ascospores. Laboulbeniomyceta are minute parasites of insects.

The phylum Basidiomycota include the fleshy fungi, mushrooms and puffballs, and the plant pathogens that cause rusts and smuts. They reproduce by a basidiospore, which arises from specialized fleshy cells called the basidia and is pinched off from the stalks (Figure XIV-6).

The phylum Deuteromycota or Fungi Imperfecti is a taxonomic dumping ground for those yeasts and molds in which no sexual reproduction has been observed. Included in this class are pathogenic fungi such as the genus *Trichophyton*, the cause of athlete's foot, and the species *Candida albicans* (Color Plate 23), the cause of an infection of the throat known as *thrush*.

65. *Culturing and Preparing Molds for Identification and Study*

The cultivation and observation of molds growing on food is facilitated by Henrici slides. For this exercise you will be asked in advance to keep a piece of bread or fruit in a plastic bag at room temperature for a week or so. You will also be given some known cultures of molds for comparison.

PROCEDURE

Preparing a Henrici slide (Figure 65-1)

1. Aseptically take a drop of melted Sabouraud's dextrose agar and place it on a sterile microscope slide.

Figure 65-1 Preparation of a Henrici slide.

2. After the drop cools and solidifies, cut a straight edge on one side of the agar with a sterile loop or an alcohol-sterilized microspatula.

3. With a sterile loop, transfer a small bit of mycelium or spores from your moldy food, or from cultures given to you by your instructor, to the cut edge of the Sabouraud's medium on the slide.

4. With an alcohol-sterilized forceps, pick up a sterile cover glass and place it on top of the inoculated medium. Slightly press the cover glass down on the medium with the forceps.

5. Transfer the slide to a sterilized moist chamber (a petri dish that contains a piece of filter paper on which is a bent glass rod).

6. Add 0.5 ml of sterile tap water to the filter paper and then place the petri dish in a sealable plastic bag and seal it.

7. Incubate the plate at room temperature (25°C) until the mold has grown to the extent wanted.

Examining the slide cultures

1. Remove the slide from the moist chamber and place it on the stage of your microscope.

2. With a low-power (10×) objective, find the edge of the mycelium. Note the characteristics of the hyphae and the reproductive structures.

3. Carefully switch to the high dry objective if you are unsure of the characteristics of the mycelium (septate or aseptate) or the reproductive structures.

4. Make drawings on your report sheet of the vegetative mycelium and as many of the characteristic reproductive structures as you can observe.

5. Try to identify your unknown as a member of one of the following common genera:

(a) Mycelium aseptate

(i) Sporangiophores arise from long, arching stolons opposite rhizoids. *Rhizopus* (Figure 65-2*A* and Color Plate 20)

(ii) Sporangiophores arise singly from mycelium at any point. All branches terminate in sporangia. Chlamydospores are frequently found in the mycelium. *Mucor* (Figure 65-2*B*)

(b) Mycelium septate

(i) The apex of the conidiophore is swollen into a vesicle from which arise bottle-shaped cells, sterigmata, which bear chains of globose conidiospores. *Aspergillus* (Figure 65-2*L*, also Figures XV-1 and XV-5 and Color Plate 22)

(ii) Conidiophores bear branched sterigmata, each with chains of conidiospores. The pattern of branching (e.g., symmetric or asymmetric and number of branches) is a characteristic used for identifying species. *Penicillium* (Figures 65-2*C*–*G* and Color Plate 22)

(iii) Sterigmata are longer, more tubular, and more widely bent away from the axis of the conidiophore than in *Penicillium*. *Paecilomyces* (Figure 65-2*H*).

(iv) Oval conidiospores are held together in clumps or balls by a gelatinous matrix. *Gliocladium* (Figure 65-2*I*).

(v) Dark brown conidia are muriform (like bricks in a wall), having both transverse and longitudinal septa, and are produced in chains from the apex of a short simple conidiophore. *Alternaria* (Figure 65-2*K*).

(vi) Conidiophores arise as upright branches from the mycelium and become bent at the apex. Multiseptate, thick-walled, cigar-shaped conidiospores are clustered at or near the bent apex. *Helminthosporium* (Figure 65-2*J*).

QUESTIONS

1. The generic names of many molds are derived from certain morphological features. What are the derivations of the names *Penicillium*, *Aspergillus*, and *Rhizopus*?

2. If you had a mixed culture of a mold and a bacterium, what selective plating procedure could you use to isolate each in a pure culture?

3. Where in the life cycle do plus and minus strains arise?

Figure 65-2 Morphology of some common molds. *A*: *Rhizopus nigricans.* *B*: *Mucor hiemalis* *C*: *Penicillium* sp. *D*: Patterns of Sterigmata and Metulae in different species of *Penicillium*. *E*: *Penicillium expansum*. *F*: *P. roqueforti*. *G*: *P. cambemberti*. *H*: *Paecilomyces*. *I*: *Gliocladium*. *J*: *Helminthosporium* sp. *K*: *Alternaria* sp. *L*: *Aspergillus*. Some of the sterigmata and conidiospores are removed to show the vesicle to which the former are attached (see also Figures XV-1 and XV-5).

Culturing and Preparing Molds for Identification and Study

Name _____

Desk No. _____

Draw and label the structures you observed in the slide cultures. Based upon your observations, make tentative identifications of the cultures.

66. *Zygospore Formation in* Rhizopus nigricans

Conjugation takes place in the Zygomycota if conditions are favorable. Zygospore formation is induced hormonally. Compatible plus and minus strains produce precursor molecules that are converted into hormones.

PROCEDURE

1. Prepare a potato-agar plate. Inoculate one side of the plate with the + strain of *R. nigricans* and the opposite side with the − strain. Label the bottom of the plate to identify the + and − sides.
2. Incubate the inoculated plate for 7 days at 30°C (Figure 66-1 and Color Plate 20).
3. Examine the growth and zygospore formation under the 16-mm and 4-mm objectives. Illustrate on your report sheet.

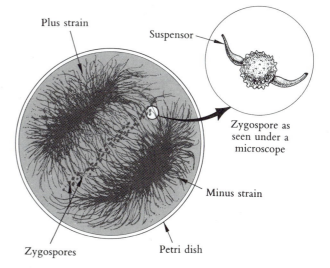

Plus strain

Suspensor

Zygospore as seen under a microscope

Minus strain

Zygospores   Petri dish

Figure 66-1   Zygospore formation by two strains of *Rhizopus.*

QUESTIONS

1. How would you go about demonstrating experimentally that it is a hormone that induces zygophore formation?

2. How would you determine the sexual polarity of a new isolate of *Rhizopus nigricans*?
3. Were there any sporangia formed on your plates? Where?

Zygospore Formation in
Rhizopus nigricans

Name _____

Desk No. _____

Draw a diagram of a zygosporangium of *Rhizopus nigricans* and the mycelial structures associated with it.

67. *Morphology and Reproduction of Yeasts*

Most yeasts reproduce mainly by means of the asexual process of budding, in which an outgrowth develops on the parent cell and eventually separates as a daughter cell. Sexual reproduction in some of the "true" yeasts (distinguished by the production of sexual spores) takes place by the conjugation of two ascospores or two somatic cells that assume the function of copulating gametangia, followed by the development of a diploid vegetative cell. Then the resulting nucleus divides into four (or fewer) nuclei. Each nucleus matures into an ascospore as it becomes encased in reserve material surrounded by a spore coat. The ascospores are retained in the parent cell, the ascus, which eventually breaks open to liberate them. In suitable growth conditions, different strains conjugate, germinate into mature vegetative cells, and the cycle is repeated (Figure 67-1).

This exercise illustrates the morphology and the mode of asexual and sexual reproduction of yeasts.

The yeasts given to you for this laboratory experiment have been grown on three types of media. These yeasts are the following:

1. *Saccharomyces* grown on eugon agar or Sabouraud's agar for general study of morphology.
2. *Saccharomyces* grown on glucose–acetate agar for 2–3 weeks to stimulate ascospore formation.
3. *Candida albicans* (opportunist and pathogen) grown on cornmeal agar to stimulate chlamydospore production (Color Plate 23).

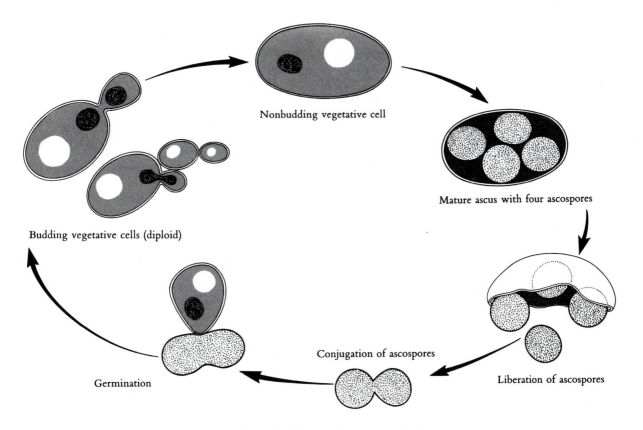

Nonbudding vegetative cell

Mature ascus with four ascospores

Budding vegetative cells (diploid)

Germination

Conjugation of ascospores

Liberation of ascospores

Figure 67-1 Life cycle of a typical yeast, *Saccharomyces cerevisiae.*

OBSERVATION

Morphology: Vegetative Cell

Prepare a wet mount of culture #1 by suspending a *small* loopful of this yeast in an iodine–water mixture (1 drop of water and 1 drop of Gram's iodine). Observe *under oil* and make sure you can identify buds, cell walls, cell membranes, glycogen vacuoles (large reddish-brown structures within the cells), and nuclei. Sketch and estimate the size of an average yeast cell.

PROCEDURE

Sexual Reproduction by Ascospore

1. Prepare an air-dried smear of culture #2 from the glucose–acetate agar by mixing a small loopful of the yeast with 2–3 loopfuls of water. DO NOT HEAT FIX. Flood entire smear with malachite green and after 5 minutes rinse dye from slide. Counterstain with safranin and after 1 minute rinse off dye. Blot dry (do not rub) and observe under oil immersion. Vegetative yeast cells will appear red, and ascospores (significantly smaller than the vegetative yeast cells) will appear green. See Figure 67-2 for description of asexual reproduction (budding) and sexual reproduction (ascospore formation).

2. Prepare a wet mount of culture #3 by suspending a loopful of the yeast grown on the cornmeal agar in two drops of water. Cover with a coverglass. Observe under the high dry objective. Filaments and chlamydospores in clusters along the mycelium are easily recognized. The budding round cells are blastospores (Figure 67-2). Ascospores are not usually found on cornmeal agar preparations.

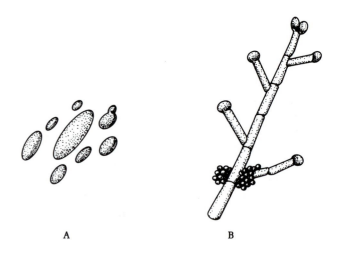

A   B

Figure 67-2 *Candida albicans.*
A: Direct mount from liquid culture showing some of the variations in the sizes of cells.
B: Filaments and chlamydospores on cornmeal agar.

3. Prepare a wet mount of a culture grown in liquid medium. Observe the variation in sizes and shapes of the yeasts (Figure 67-2). Illustrate your observations on the report sheet.

QUESTIONS

1. From your observations would you conclude that spore formation occurs more frequently among the molds or the yeasts?
2. Certain microbial forms bridge the morphological gap between molds and yeasts. What are they and what characteristics do they share with each group?
3. What does the term *Saccharomyces* mean?

REPORT 67

Morphology and Reproduction of Yeasts

Name _____

Desk No. _____

Make drawings of typical yeast structures and label.

Soil Microbiology

A handful of soil contains a world of microorganisms, vast numbers of many kinds. The conditions in soils can vary so much that the dominant microflora may be entirely different in two samples that are feet, or even inches, apart.

In nature, the microflora of a habitat changes constantly in response to the changing environment. In autumn, the falling of ripened, fermenting fruits provides habitats for populations of yeasts, which are soon followed by populations of *Acetobacter* that obtain energy by oxidizing the alcohol formed by the yeasts. Among the far-reaching consequences of the liming of a field are the likely development of *Azotobacter*, which usually does not grow in a habitat below pH 6, and the inhibition of such acid-tolerant species as *Thiobacillus thiooxidans*, the sulfur-oxidizing autotroph that flourishes at pH 2 or lower. A sudden shower, resulting in an accumulation of water in low areas of the field, makes a quagmire that prevents free diffusion of gases, and the thriving aerobic flora of bacilli soon gives way to fermentative or anaerobic species. Anaerobic processes such as denitrification by *Pseudomonas* replace the aerobic processes such as nitrification that were formerly carried on by the aerobic autotrophs *Nitrosomonas* and *Nitrobacter*. The fluctuations in temperature from a hot August midday to a cool night alternately favor different groups of microorganisms. At certain times, after periods that are beneficial to the fungi, sporangia ripen and burst, and spores are scattered by the wind in enormous numbers over the field—making, for the time being, a predominating surface flora of mold. Behind this readily apparent scene of shifting population, a myriad of other organisms —protozoa, actinomycetes, symbiotic nitrogen fixers, cellulose decomposers, and others—are engaged in chemical activities so enmeshed and interdependent as to defy any but the most naive estimations. As the environment changes, each group of microorganisms comes into its own, perhaps not as a dominant population, but as a minority favored by chance for a short time.

It should be emphasized that the biochemistry of the various groups of microbes in soil is not well understood. Until a decade ago, it was thought that nitrogen fixation was carried out by three genera, *Rhizobium, Azotobacter*, and *Clostridium*. Although they remain the eminent nitrogen-fixing genera, it can be shown through use of isotopic nitrogen that a variety of bacteria and cyanobacteria are capable of fixing atmospheric nitrogen. An analogous situation may exist concerning nitrification, which now appears to be carried out by only two genera of autotrophic microorganisms; future work may demonstrate that participation in nitrification is more widespread than this.

68. *Microbial Populations of Soil*

Many of the microorganisms in soil, protozoa and algae, cannot be grown by the usual techniques of the microbiologist. Even so, the microbial population, comprising chiefly bacteria, fungi, and actinomycetes, and running to many millions of individuals per gram of good soil, can be demonstrated in the laboratory.

PROCEDURE

Two samples of soil are provided: a rich garden soil and a sandy loam. Plate out each sample as indicated.

1. Weigh out 1 g of soil and place it directly into a 99 ml dilution blank. Let it stand for 10–15 minutes and shake it well.
2. Prepare additional dilutions of the sample (as in Exercise 17) and plate out dilutions of 10^{-4}, 10^{-5}, and 10^{-6}, using tryptone–yeast-extract agar.
3. Incubate the plates at 30°C for 4 days; after counting, incubate for another week.
4. Count the number of colonies and calculate the number of bacteria per gram of each soil sample.
5. Make gram stains of five colony types.
6. Record your findings on the report sheet.

QUESTIONS

1. What general types of soil-dwelling bacteria will not appear in the results of this plating?
2. How would you modify this procedure to get an estimate of the number of spore formers per gram of soil?
3. What steps would you take to get an estimate of the fungal population, not complicated by bacterial colonies?

Microbial Populations of Soil

Name _____

Desk No. _____

Sample	Dilution counted	Number of colonies	Plate count per gram of soil
Rich garden soil			
Sandy loam			

Make drawings of bacterial types from five different colonies. Make a guess about the identity of each.

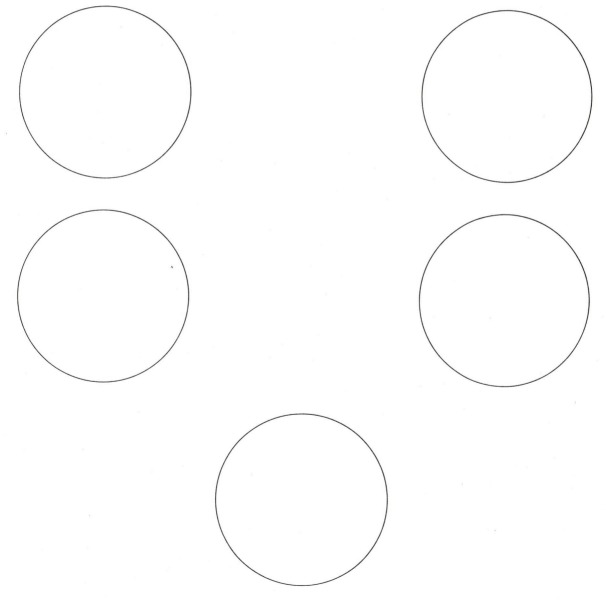

69. *Nitrogen Cycle*

Although nitrogen is an essential element for all life, the nature of the nitrogen compounds that satisfy this nutritional requirement varies. Life depends on a continual replenishment of nitrogen compounds in a range of states of oxidation–reduction. Microorganisms play a major role in these transformations, collectively termed the **nitrogen cycle**.

Plants depend on ammonia or nitrates as their nitrogen sources, and animals depend on organic nitrogen; microorganisms vary widely in their requirements. This diversity in nitrogen metabolism provides the essential steps in the nitrogen cycle. In this exercise we consider the transformations that represent a major part of this cycle.

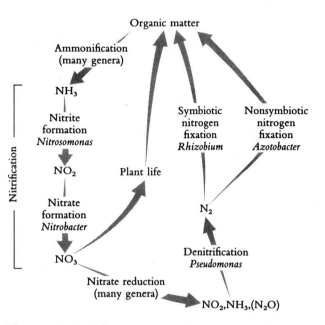

Figure 69-1 The nitrogen cycle. The bacterial genera named are important examples of those carrying out the reactions.

AMMONIFICATION

The organic nitrogen compounds in nature — animal, plant, and microbial protein, as well as wastes such as urea — are subject to dissimilation by a variety of heterotrophic microorganisms, which yield ammonia and various other end products (Figure 69-1).

PROCEDURE

1. Treat five tubes of a 4% peptone solution, each in a different way, as follows:
   (a) Leave one as a sterile control.
   (b) Inoculate one with a loopful of soil.
   (c) Inoculate one with *Bacillus cereus*.
   (d) Inoculate one with *Pseudomonas fluorescens*.
   (e) Inoculate one with *Proteus vulgaris*.
2. Incubate all five tubes at 30°C for 48 hours.
3. Test for the presence of ammonia: Place a drop of Nessler's reagent on a spot plate; with a glass rod or a pipet, add a drop of the substance being tested. The presence of ammonia is shown by a yellow color.
4. Record your findings on the report sheet.

NITRIFICATION

The ammonia produced from the degradation of proteins, amino acids, and other organic forms of nitrogen can be assimilated by microorganisms into cell protein or oxidized, first to nitrites, and then to nitrates by nitrifying bacteria. The oxidation, called **nitrification**, is carried out stepwise by two genera of autotrophic bacteria: the genus *Nitrosomonas* oxidizes ammonia to nitrites, and the genus *Nitrobacter* oxidizes nitrites to nitrates (Figure 69-1).

PROCEDURE

Nitrite Formation

1. Inoculate the nitrite-formation medium*
   with about 0.1 g of soil (with a neutral or

*The nitrite-formation medium must be free from nitrogen in all forms except as an ammonium salt.

slightly alkaline pH) and incubate at 25–30°C.

2. At intervals of one week, test the culture for the presence of nitrites: Mix three drops of Trommsdorf's solution with one drop of sulfuric acid (one part concentrated H_2SO_4 to three parts H_2O) in a spot plate. Add a drop of culture from a glass rod or pipet and mix. The appearance of an intense blue-black color shows the presence of nitrites. (*Note*: A glass rod is used because wire loops may give a false reaction.)

3. Test the culture for the presence of ammonia (using Nessler's reagent) once a week. After prolonged incubation, this test should become negative as a result of the complete oxidation of ammonia to nitrites.

4. Each week, after suspending the sediment, make gram stains and examine for the presence of typical forms.

Nitrate Formation

1. Inoculate the nitrate-formation medium with about 0.1 g of soil. Incubate at 30°C.

2. Test the flask each week with Trommsdorf's reagent until the test turns negative for nitrites. This is important because the diphenylamine reagent used to test for nitrite gives a positive test for both nitrites and nitrates.

3. On a spot plate put one drop of diphenylamine reagent and two drops of concentrated H_2SO_4. Add a drop of the culture being examined. A dark blue-black color shows the presence of nitrates.

4. Make gram stains each week after suspending sediment and examine for the presence of typical forms (see Figure 69-1).

DENITRIFICATION

The reduction of nitrates is brought about by a number of soil-dwelling species. Nitrites, ammonia, nitrous oxide, and nitrogen gas may be among the end products of **nitrate reduction**. If reduction to nitrogen gas is complete, the process is called **denitrification**. Some species actively carry out ammonification under aerobic conditions, and under anaerobic conditions they catalyze denitrification. The nitrates serve as H-acceptors for these

species in the absence of oxygen in a reaction termed **anaerobic respiration** (Figure 69-1).

PROCEDURE

1. Inoculate one tube of nitrate broth containing a Durham tube with about 0.1 g of soil and another with *Pseudomonas aeruginosa*.

2. Inoculate one tube of nitrate-free broth containing a Durham tube with *P. aeruginosa* and a second with 0.1 g of soil.

3. Incubate the four tubes for 1 week at 30°C.

4. Observe for gas formation.

5. Add 1 ml of α-naphthylamine reagent (*Caution*: do *not* use a mouth-drawn pipet) and 1 ml of sulfanilic acid reagent to the culture tubes after incubation. A red color within 30 seconds indicates that nitrites are present.

NITROGEN FIXATION

A vastly important role of microorganisms in nature is their ability, either in association with leguminous plants or alone, to use atmospheric nitrogen as a source of their cell nitrogen. The ability to carry out **nitrogen fixation** appears to be limited to a relatively few genera.

Symbiotic nitrogen fixation. The roots of most of the leguminous plants, such as peas, beans, and clovers, bear numerous nodules or tumorlike growths (Figure 69-2). These nodules form as a result of an infection of the root hairs by the bacterial genus *Rhizobium*. The infecting *Rhizobium* cells are symbiotic on the legume, obtaining sources of energy for growth from the host and growing to large populations in the nodule. During this growth, the bacteria fix gaseous nitrogen from the air and make it available to the plant. Thus, the bacteria and the plants exist in a symbiotic, or mutually beneficial, relationship (see Figure 69-1).

Nonsymbiotic nitrogen fixation. In contrast to nitrogen fixation by the genus *Rhizobium*, several other species of microorganisms fix nitrogen while living independently in the soil. Foremost of these is the bacterial genus *Azotobacter*, which oxidizes organic wastes in the soil as a source of energy and fixes atmospheric nitrogen as a source of cell nitrogen. The blue-green bacteria, the clostridia, the purple

Figure 69-2 Root nodules on clover plant containing nitrogen-fixing bacteria.

nonsulfur photosynthetic bacteria, and several gram-negative forms also use gaseous nitrogen nonsymbiotically.

The bacteria that fix free nitrogen from the air can be grown on a medium that contains no nitrogen, but only inorganic salts and a simple carbohydrate, from which they derive energy.

PROCEDURE

Symbiotic Nitrogen Fixation

1. Select a nodule from the root structure of the legume provided and crush it in a drop of water between two slides.
2. From this preparation transfer a loopful to a clean slide, make a smear, and stain it with crystal violet.
3. Examine under the oil-immersion lens and look especially for irregularly shaped cells such as pear- , club- , Y- , and T-shaped individuals. These are "bacteroid" shapes and appear in preparations from nodules but not in cultures from artificial media, in which the organisms grow as rods.

4. Prepare and stain a smear of the pure culture of *Rhizobium* grown on glucose yeast-extract agar.

Nonsymbiotic Nitrogen Fixation

1. Inoculate a flask containing a shallow layer of mannitol-salts medium with about 1 g of fertile soil.* Incubate at 30°C for a few days.
2. When surface growth has developed, make stained preparations and examine them for large rods or coccoid cells embedded in slimy material.
3. Examine the mannitol-agar culture of *Azotobacter chroococcum*. Note the glistening, slimy growth. Prepare a stained preparation and compare the appearance of the laboratory-grown strain with the fresh soil strain.

QUESTIONS

1. What groups of organisms, in addition to the bacteria, carry out ammonification in the soil?
2. From what chemical grouping in protein does NH_3 originate?
3. Will nitrifying bacteria grow on organic media?
4. What is the source of energy for *Nitrosomonas*? For *Nitrobacter*?
5. What environmental conditions would bring about an increase in denitrification and a decrease in nitrification in the soil?
6. How does nitrate reduction differ from denitrification?
7. Why might you get gas formation in the tube of nitrate-free media inoculated with soil?
8. How do you account for the presence of other morphological types of bacteria that are often observed in stained preparations of fresh *Azotobacter* growth from the nitrogen-free medium?
9. What would you expect to develop if glucose were used instead of mannitol in the medium?

---

*Mannitol, although a ready energy source for *Azotobacter*, is not readily utilized by many competing soil organisms.

Nitrogen Cycle

Name _____

Desk No. _____

AMMONIFICATION

Culture	NH_3
Soil	
Bacillus cereus	
Pseudomonas fluorescens	
Proteus vulgaris	

NITRIFICATION: NITRITE FORMATION

Time	NH_3	NO_2	Morphology and gram reaction
1st week			
2nd week			
3rd week			

NITRIFICATION: NITRATE FORMATION

Time	NO_2	NO_3	Morphology and gram reaction
1st week			
2nd week			
3rd week			

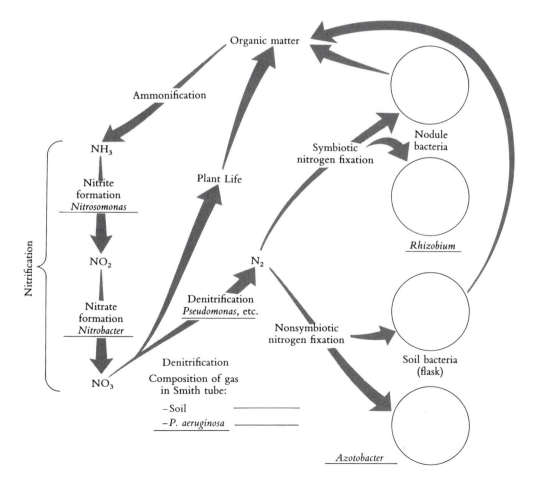

Medical Microbiology and Immunology

Medical microbiology has received the most research attention over the years because it is concerned directly with human welfare. The isolation and identification of pathogenic organisms preceded the development of effective control measures for the diseases.

Immunology, a science closely allied to medical microbiology, deals with the development of immunity to disease. **Serology**, another allied science, deals with the biological and chemical properties of blood serum; many serological techniques are important in medical microbiology.

This section introduces studies of the parasites and pathogens of the human mouth and throat, methods to detect other types of human pathogens, and some of the techniques of immunology and serology.

70. *Normal Mouth, Nose, and Throat Flora*

The mouth, nose, and throat of a healthy person contain many bacteria, most of them harmless. Some are potential pathogens but virulent organisms can be present without producing disease, even though the same species may be the cause of disease in other persons. (see Color Plate 24.)

Organisms that can be isolated from healthy persons include *Streptococcus mutans*, *S. sanguis*, *S. mitior*, *S. milleri*, *S. salivarius*, and anaerobic streptococci; *Micrococcus mucilagenosus*, *Actinomyces israelii*, *A. viscosus*, *A. odontolyticus*, *Lactobacillus casei*, *L. acidophilus*, *Arachnia propionica*, *Propionibacterium* spp., *Rothia dentocariosa*, *Bacterionema matruchotii*, *Eubacterium saburreum*, *E. alactolyticum*, *Bifidobacterium eriksonii*, *B. dentium*, *Veillonella alcalescens*, *V. parvula*, *Neisseria* spp., *Branhamella* sp.; *Staphylococcus aureus*, *Staphylococcus* spp., *Haemophilus influenzae*, *H. parainfluenzae*, *H. segnis*, *H. aphrophilus*, *H. par-* aaphrophilus, *Eikenella corrodens*, *Bacterioides melaninogenicus*, *B. oralis*, *B. corrodens*, *B. ruminicola*, *B. ochraceus*, *Selenomonas sputigena*, *Vibrio succinogenes*, *Fusobacterium nucleatum*, *Simonsiella* spp., *Leptotricha buccalis*, oral spirochetes, the protozoa *Trichomonas tenax* and *Endamoeba gingivalis*, and yeasts.

An enriched medium, in which red blood corpuscles are included in the agar, is valuable for identifying certain pathogens. Blood agar is usually employed in streak plates, but it can be used in pour plates or tubes and slanted for growth and storage of fastidious organisms. The reaction given by a bacterial species can vary with the source of blood. Horse blood is preferred, but other types of blood can be used. The medium to which the blood is added has a high percentage of NaCl (0.5%) to prevent spontaneous lysis of the red blood cells.

In this experiment you will swab your mouth,

nose, and throat, and inoculate a blood agar plate and an EMB plate. You will choose one colony type from each plate and follow procedures to identify them. First you will make sure your isolates are pure (axenic), then you will gram stain them. Most probably your gram positive isolate will be either a *Streptococcus* or *Staphylococcus*. You will recognize which you have isolated by their grouping in the gram stain. A different bank of tests is used to identify species of either genus. You will use one of the miniaturized multitest systems (e.g., API 20E or Enterotube II) to identify the gram negative isolate.

PROCEDURE

A. Preparing Blood Agar Plates

1. Melt a tube (12 ml) of blood-base agar and cool it to 43–45°C.
2. Temper a tube of sterile blood in the water bath at 43–45°C.
3. Transfer aseptically 1.0 ml of blood to the cooled tube of agar and mix with the pipet.
4. Pour the mixture into a petri plate and rotate to complete the mixing. If bubbles appear on the surface of the agar, break them by inverting a burner and playing the flame *lightly* over the surface before the agar solidifies.

B. Obtaining a Throat, Mouth, or Nose Culture

1. Obtain two swubes, two blood agar plates, an EMB plate, and a tube of saline.
2. Ask your laboratory partner to help you obtain samples from your nose or throat.
3. Moisten the swube in sterile saline (so it won't irritate tissues during the sampling; it also enhances the pickup of organisms).
4. After moistening one of the swubes, swirl it inside your nostril or in the back of your throat (Figure 70-1).
5. Swirl and swab the swube on a blood agar plate, following the technique demonstrated by your laboratory instructor.
6. Moisten a second swube and take a sample from inside your mouth (tongue, gums, roof, etc.).
7. Swirl and swab this swube on the EMB plate and on a blood agar plate, using the same technique as before.
8. Incubate the plates in a 37°C incubator for 18–24 hr (Color Plate 25).

*A swube is a sterile, disposable swab in a test tube (Fischer Scientific).

Swube   Saline

Figure 70-1 Making a nasal swab.

Reactions on Blood Agar

Different organisms that are grown in blood agar may exhibit various reactions (Color Plate 26). Hemolytic organisms exhibit **hemolysis**, also called B-hemolysis. A zone of varying diameter (from less than 1 mm to 2–3 cm) develops around the colony. In this zone, the red blood cells have disintegrated, and the red color has disappeared. Hemolysis is caused by hemolysins, hemolytic agents released by the bacterial cells. Nonhemolytic organisms exhibit *no hemolysis*, called γ-reaction, in which there is no noticeable effect of bacterial cells growing on the red blood cells.

Certain lactic acid bacteria such as *Streptococcus pneumoniae* and *Streptococcus mitis* exhibit greening, also called α-reaction, which is the appearance of a greenish tint in areas around the colonies, sometimes with partial disintegration of the individual red blood cells. Greening is associated with the production of hydrogen peroxide.

C. After 18–24 hours of incubation

1. Retrieve your plate from the incubator.
2. Note the color, characteristics, and distribution of the various kinds of colonies on your plate. Your instructor will go over them with you next period.
3. Store your plates in a refrigerator.

QUESTIONS

1. Why doesn't the blood used in this experiment coagulate prior to use?
2. What is defibrinated blood? Citrated or oxalated blood?
3. Media used for the demonstration of blood reactions should be low in fermentable carbohydrates. Explain.

D. Next Laboratory Period

1. Choose one colony type from each plate (EMB and blood agar).
2. Make gram stains of cells from your chosen type colonies. Make sure you have chosen one gram (+) (from blood agar) and one gram (−) (from the EMB).
3. Streak your chosen colony type on the same type of medium that you used for the initial isolation.
4. Incubate the plates in the designated area in the 37°C incubator. Store your original plates in a refrigerator. Make sure both are properly labeled.

E. After 18–24 hours of incubation

1. Remove your two plates from the incubator.
2. Examine them to see if you have isolated only one colony type and that you have cleanly isolated colonies.
3. If you have, store your plates in the refrigerator. If you haven't, restreak and examine 24 hours later.

Identifying the Gram-Negative Isolate

If you haven't used it previously, your laboratory instructor will demonstrate the use of the Enterotube II miniaturized multitest system for the rapid identification of Enterobacteriaceae.

1. Pick a colony from your restreaked EMB plate and read in your laboratory manual about the Enterotube II system (Exercise 41).
2. Follow the directions in Exercise 41.
   Step 6: Break the wire when tip is in H₂S/indole compartment.
   Step 8: Remove blue strip
   Step 9: Slide clear band over glucose compartment.
3. Incubate your tube flat in a 37°C incubator.

After 18–24 hours of incubation

1. *After 18–24 hours* retrieve your tube and follow the directions for the second laboratory period of Exercise 41.
2. Store your tubes in the refrigerator until the class discussion next period.

Identifying a Gram-Positive Isolate: Staphylococcus

If you isolated a *Staphylococcus* (determined by the gram stain you did last period), read the rest of this exercise.

You will use the STAPH-IDENT System to identify your *Staphylococcus*. *Before* setting up this identification technique, take into consideration that it must be completed at the end of 5 hours. Arrange a mutually convenient time. Your laboratory instructor will give you the parameters.

API STAPH-IDENT System (Figure 70-2)

1. Prepare a suspension of your unknown *Staphylococcus* by picking one or two colonies from your trypticase-soy blood agar plate with a loop and mixing them with 2 ml of 0.85% sterile saline (pH 5.5–7.0) in the tube provided.
2. Mix the suspension thoroughly and then compare it to a #3 McFarland (BaSO₄) turbidity standard. If needed, pick additional colonies to increase turbidity or dilute with saline until the suspension density is approximately equal to the standard.
3. Remove an API STAPH-IDENT strip from the sealed envelope and place it into an API incubation tray to which 5 ml of tap water has been added (Figure 70-2A–C).
4. With a pipet transfer approximately 80 μl (2 or 3 drops) of your bacterial suspension to fill each microcupule in the test strip.
5. After inoculation, place the plastic lid on the tray and incubate the strip for 5 hours at 35–37°C.

After 5 hours of incubation

6. After 5 hours, record the results of all the tests except 10, the NGP (see Table 70-1).
7. After the STAPH-IDENT reagent has come to room temperature, add 1 or 2 drops to the NGP microcupule. Allow 30 seconds for color development and record results.
8. Construct a four-digit profile in the following way:
   (a) The 10 biochemical tests are divided into four groups:

PHS	MNE	SAL	NGP
URE	MAN	GLC	
GLS	TRE	ARG	

   (b) Assign a numerical value to positive reactions. The value depends on the location of the reaction in its group. For example:

   A value of ONE for the first biochemical in each group (i.e., PHS, MNE . . .)
   A value of TWO for the second biochemical in each group (i.e., URE, MAN . . .)
   A value of FOUR for the third biochemical in each group (i.e., GLS, TRE . . .)

A four-digit number is obtained from the totals of the values of each group. The four-digit number is used as the input for the computer.

Example: 7700 — *Staphylococcus aureus*

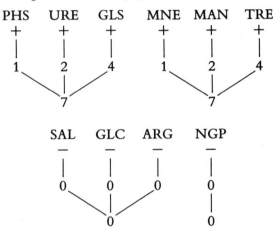

Note: Each number represents only one combination; therefore, for each four-digit number, there will be only one corresponding profile.

Profile numbers with more than one identification choice require additional testing before a final identification can be determined. The tests that are recommended for the separation of the species listed, along with the expected results, are printed to the right of each identification choice.

9. Compare the results with Table 70-2.
10. After all reactions have been recorded on the report sheet and a satisfactory identification has been made, the entire incubation unit must be autoclaved, incinerated, or immersed in a germicide before disposal.

ADDITIONAL TESTS FOR *STAPHYLOCOCCUS*

Heat-Stable DNase Test

In addition to coagulase production, (below) most infectious and food-poisoning strains of *Staphylococcus* produce a thermostable *nuclease*, an enzyme that depolymerizes DNA.

To demonstrate the presence of this enzyme, agar that contains dissolved DNA and a dye, methyl green O, is exposed to a culture of the microorganism being examined. If nuclease is

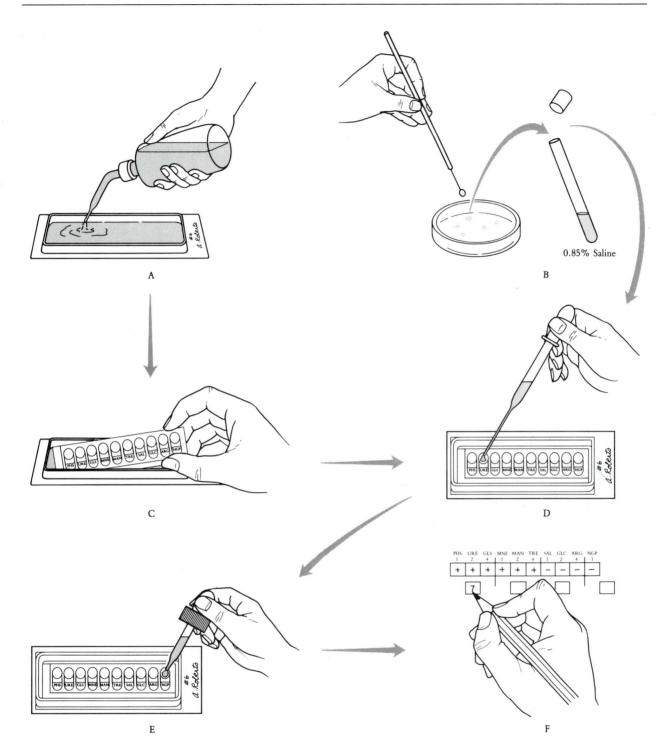

Figure 70-2 Flow chart of the API STAPH-IDENT procedure. *A*: Place 5 ml of tap water into bottom of tray. *B*: With a loop, make a saline suspension of *Staphylococcus* that is equal in turbidity to a McFarland No. 3 barium sulfate standard. *C*: Place a STAPH-IDENT test strip in the bottom of the moistened tray. *D*: Inoculate the bacterial suspension into each of the 10 microcupules and cover tray with lid. *E*: After incubation at 37°C for 5 hours, add 1 or 2 drops of STAPH-IDENT reagent to the tenth microcupule. *F*: Record results and add up positive values in each test group.

Table 70-1 Interpretation of the Results of the API STAPH-IDENT System

Microcupule Numbers	Test	Positive	Negative	Comments and References	Chemical/Physical Principles
1	PHS Phosphatase	Yellow	Clear or straw colored	Positive only if significant color development has occurred.	Hydrolysis of p-nitrophenyl-phosphate, disodium salt, by alkaline phosphatase releases yellow p-nitrophenol from the colorless substrate.
2	URE Urea utilization	Purple to red-orange	Yellow or yellow-orange	Phenol red has been added to the urea formulation for detection of alkaline end products resulting from urea utilization.	Urease releases ammonia from urea; ammonia causes the pH to rise and changes the indicator from yellow to red.
3	GLS Glycosidase	Yellow	Clear or straw-colored	Positive only if significant color development has occurred.	Hydrolysis of p-nitrophenyl-β-D-glucopyranoside by β-glucosidase releases yellow p-nitrophenol from the colorless substrate.
4 5 6 7	MNE MAN TRE SAL Utilization of mannose mannitol trehalose or salicine	Yellow or yellow-orange	Red or orange	Cresol red has been added to each carbohydrate to detect acid production if the carbohydrate is utilized.	Utilization of carbohydrate results in acid formation. Indicator changes from red to yellow.
8	GLC β-glucuronidase	Yellow	Clear or colorless	Positive only if significant color development has occurred.	Hydrolysis of p-nitrophenyl-β-D-glucuronide by β-glucuronidase releases yellow p-nitrophenol from the colorless substrate.
9	ARG Arginine utilization	Purple to red	Yellow or yellow-orange	Phenol red has been added to the arginine formulation to detect alkaline end products resulting from arginine utilization.	Utilization of arginine produces alkaline end products which change the indicator from yellow to red.
10	NGP β-galactosidase	Add 1-2 drops of STAPH-IDENT reagent Plum-purple (mauve)	Yellow or colorless	Color development begins within 30 seconds of reagent addition.	Hydrolysis of 2-naphthyl-β-D-galactopyranoside by β-galactosidase releases free β-naphthol which complexes with STAPH-IDENT Reagent to produce a plum-purple (mauve) color.

Table 70-2 STAPH-IDENT Profile Numbers and Identifications

Profile	Identification		Profile	Identification	
0 040	Staph capitis		4 700	Staph aureus	COAG+
0 060	Staph haemolyticus			Staph sciuri	COAG−
0 100	Staph capitis		4 710	Staph sciuri	
0 140	Staph capitis		5 040	Staph epidermidis	
0 200	Staph cohnii		5 200	Staph sciuri	
0 240	Staph capitis		5 210	Staph sciuri	
0 300	Staph capitis		5 300	Staph aureus	COAG+
0 340	Staph capitis			Staph sciuri	COAG−
0 440	Staph haemolyticus		5 310	Staph sciuri	
0 460	Staph haemolyticus		5 600	Staph sciuri	
0 600	Staph cohnii		5 610	Staph sciuri	
0 620	Staph haemolyticus		5 700	Staph aureus	COAG+
0 640	Staph haemolyticus			Staph sciuri	COAG−
0 660	Staph haemolyticus		5 710	Staph sciuri	
1 000	Staph epidermidis		5 740	Staph aureus	
1 040	Staph epidermidis		6 001	Staph xylosus	XYL+ ARA+
1 300	Staph aureus			Staph saprophyticus	XYL− ARA−
1 540	Staph hyicus (an)		6 011	Staph xylosus	
1 560	Staph hyicus (an)		6 021	Staph xylosus	
2 000	Staph saprophyticus		6 101	Staph xylosus	
	Staph hominis		6 121	Staph xylosus	
2 001	Staph saprophyticus	NOVO R	6 221	Staph xylosus	
2 040	Staph saprophyticus	NOVO S	6 300	Staph aureus	
	Staph hominis		6 301	Staph xylosus	
2 041	Staph simulans	NOVO R	6 311	Staph xylosus	
2 061	Staph simulans	NOVO S	6 321	Staph xylosus	
2 141	Staph simulans		6 340	Staph aureus	COAG+
2 161	Staph simulans			Staph warneri	COAG−
2 201	Staph saprophyticus		6 400	Staph warneri	
2 241	Staph simulans		6 401	Staph xylosus	XYL+ ARA+
2 261	Staph simulans			Staph saprophyticus	XYL− ARA−
2 341	Staph simulans		6 421	Staph xylosus	
2 361	Staph simulans		6 460	Staph warneri	
2 400	Staph hominis		6 501	Staph xylosus	
	Staph saprophyticus		6 521	Staph xylosus	
2 401	Staph saprophyticus	NOVO S	6 600	Staph warneri	
2 421	Staph simulans	NOVO R	6 601	Staph saprophyticus	XYL− ARA−
2 441	Staph simulans			Staph xylosus	XYL+ ARA+
2 461	Staph simulans		6 611	Staph xylosus	
2 541	Staph simulans		6 621	Staph xylosus	
2 561	Staph simulans		6 700	Staph aureus	
2 601	Staph saprophyticus		6 701	Staph xylosus	
2 611	Staph saprophyticus		6 721	Staph xylosus	
2 661	Staph simulans		6 731	Staph xylosus	
2 721	Staph cohnii (ssp1)		7 000	Staph epidermidis	
2 741	Staph simulans		7 021	Staph xylosus	
2 761	Staph simulans		7 040	Staph epidermidis	
3 000	Staph epidermidis		7 141	Staph intermedius (an)	
3 040	Staph epidermidis		7 300	Staph aureus	
3 140	Staph epidermidis		7 321	Staph xylosus	
3 540	Staph hyicus (an)		7 340	Staph aureus	
3 541	Staph intermedius (an)		7 401	Staph xylosus	
3 560	Staph hyicus (an)		7 421	Staph xylosus	
3 601	Staph simulans		7 501	Staph intermedius (an)	COAG+
	Staph saprophyticus			Staph xylosus	COAG−
4 060	Staph haemolyticus	NOVO S	7 521	Staph xylosus	
4 210	Staph sciuri	NOVO R	7 541	Staph intermedius (an)	
4 310	Staph sciuri		7 560	Staph hyicus (an)	
4 420	Staph haemolyticus		7 601	Staph xylosus	
4 440	Staph haemolyticus		7 621	Staph xylosus	
4 460	Staph haemolyticus		7 631	Staph xylosus	
4 610	Staph sciuri		7 700	Staph aureus	
4 620	Staph haemolyticus		7 701	Staph xylosus	
4 660	Staph haemolyticus		7 721	Staph xylosus	
			7 740	Staph aureus	

present, bright clear zones appear where the DNA has been hydrolyzed.

1. Streak 3 DNAase methyl-green agar plates (DIFCO 0220) with cultures of *Staphylococcus aureus, Serratia marcescens,* and *E. coli.*
2. Incubate 18–24 hours at 37°C.
3. Observe for clear zones around the colonies. Clearing indicates depolymerized DNA.

Coagulase Test

The Bacto Staph Latex Test is based on the simultaneous agglutination of coagulase (clumping factor) and protein A with yellow latex particles sensitized with antigen-specific plasma proteins. Staphylococcal colonies containing clumping factor and/or protein A, when mixed with yellow latex reagent, agglutinate within 45 seconds into clumps visible to the unaided eye. Positive control is *S. aureus* (ATCC 25923). Negative control is *S. epidermidis* (ATCC 12228).

Warning: All reagents contain 0.2% sodium azide as a preservative:

Allow all reagents to equilibrate to room temperature for about 10–15 minutes after removing them from the refrigerator (2–8°C). If a sufficient number of staphylococcal colonies of the recommended size (1–3 mm) are available, test these colonies using Method A. If the number of available colonies is limited or if the colonies are small, emulsify them directly into the Bacto Staph Latex Reagent, using Method B, which has greater sensitivity. However, stringy colonies and colonies that might agglutinate in saline should not be tested by this method.

Method A

1. Place one drop of the Bacto Staph Positive Control into the first circle on the test slide.
2. Place one drop of the Bacto Staph Negative Control into the second circle on the test slide.
3. Add one drop of the Bacto Normal Saline Reagent to all remaining circles, depending on the number of staphylococcal specimens to be tested.
4. Using an inoculating needle, transfer four colonies, or more, from each suspected staphylococcal culture to a circle containing the Bacto Normal Saline Reagent and emulsify them thoroughly.

5. Add one drop of the Bacto Staph Latex Reagent to each circle.
6. Quickly mix the latex reagent with the contents of each circle, each with a fresh mixing stick.
7. Rotate the slides by hand for 45 seconds and read the results. Alternatively, place the slide(s) on a slide rotator capable of providing 100–125 rpm and rotate the slide for 45 seconds.
8. Read the results of the agglutination immediately, following these guidelines: If the latex agglutination reactions occur very rapidly, read the test results even earlier than 45 seconds. *Note:* Hold the slide at arm's length while reading. Use an indirect light source for easier reading.

Method B

1. Place one drop of the Bacto Staph Positive Control into the first circle on the test slide.
2. Place one drop of the Bacto Staph Negative Control into the second circle on the test slide.
3. Add one drop of the Bacto Staph Latex Reagent to each of these circles and to additional circles designated for testing the specimens.
4. Using an inoculating needle, quickly transfer two or three well-formed colonies from each presumptive staphylococcal culture to a specified circle and emulsify the colonies thoroughly with the latex reagent. This step should be done quickly to minimize drying of the latex on the slide.
5. Rotate the slides by hand for 45 seconds and read the results. Alternatively, place the slide(s) on a slide rotator capable of providing 100–125 rpm and rotate the slide for 45 seconds.
6. Read the results of the agglutination immediately following these guidelines: If the latex agglutination reactions occur very rapidly, read the test results earlier than 45 seconds. *Note:* Hold the slide at arm's length while reading. Use an indirect light source for easier reading.

Results

The positive test results are described in Figure 70-3.

Figure 70-3 Positive results of Bacto Staph Latex Test. *A*: Fine to grainy clumps of yellow latex beads in milky background. *B*: Small clumps of beads in milky background. *C*: Large and small clumps in slightly cloudy background. *D*: Large coarse clumps of beads in a clear background.

Negative Tests

± Smooth milky yellow suspension with a particulate appearance that cannot be identified as agglutination.
− No agglutination, smooth milky yellow suspension as with the negative control.

Interpretation of Coagulase Tests

This test is considered positive if definitive agglutination patterns of large and fine clumps are visible to the unaided eye. The results should be read in parallel with the negative control.

Limitations of the Coagulase Test

1. Some staphylococcal colonies give stronger agglutination reactions when grown on tryptic-soy agar with 5% sheep blood than on brain–heart infusion agar.
2. Very rarely, coagulase-positive staphylococci, other than *S. aureus*, and some yeasts, will yield positive test results with the Bacto Staph Latex Reagent.

3. Media with high salt concentration can cause a reduced protein A secretion from *S. aureus*. Colonies of staphylococci grown on such media may not yield significant agglutination reactions.
4. Infrequently, some staphylococcal specimens yield stringy agglutination reactions with the Bacto Staph Latex Reagent. Interpretations of such results should be based on the clarity of the background, as follows: Tests are to be considered positive with a clear background, and negative with a cloudy background.

IDENTIFYING A GRAM-POSITIVE ISOLATE: *STREPTOCOCCUS*

If you isolated only pure *Streptococcus* on your blood agar plate (determined by the gram stain you did), look over this exercise to get a general feeling for this type of miniaturized multitest system (Figure 70-4). You will use the API Rapid STREP System to identify your *Streptococcus*. *Note:* Aseptic technique must be used.

This procedure is not for use directly with clinical or other specimens. The microorganism to be identified must first be isolated as separate colonies by streaking the specimen onto appropriate culture media (e.g., Columbia Sheep blood agar) according to standard microbiological techniques.

API Rapid STREP System

1. Pick one well-isolated colony and make a suspension by homogenizing it in 0.3 ml of sterile distilled water. Flood a petri dish of Columbia Sheep blood agar with this suspension (or swab the surface of the agar with it). Incubate the plate anaerobically for 18–24 hours at 35–37°C.
2. After incubation, harvest all the culture from the petri dish with a sterile swab. Make a suspension in 2 ml of distilled water. Homogenize carefully. This suspension should be very dense, with a turbidity greater than No. 4 McFarland Barium Sulfate turbidity standard.
   Note: B-hemolytic *Streptococci* and *Enterococci* produce reasonably sized colonies after 24 hours incubation. An incubation time of 48 hours is recommended for other *Strepto-*

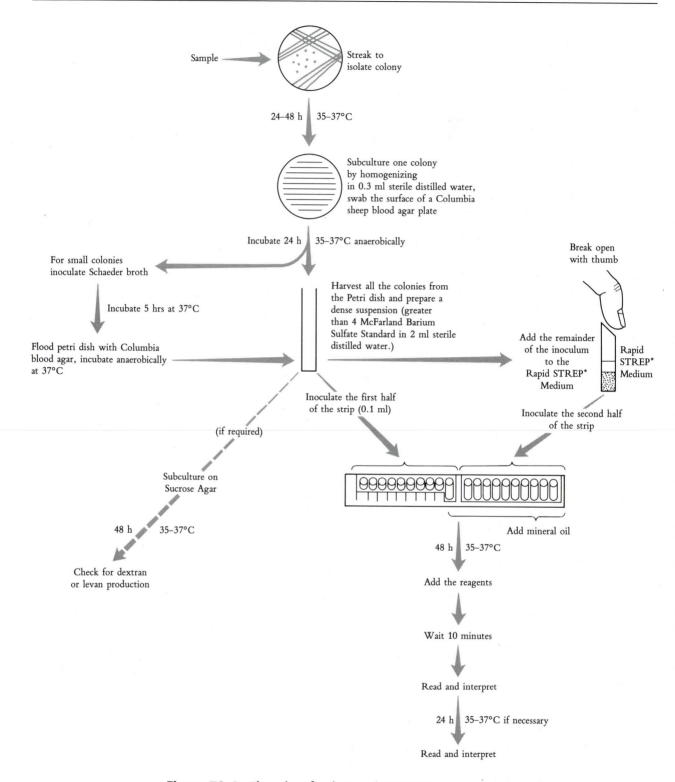

Figure 70-4 Flow chart for the Rapid-STREP diagnostic system.

cocci prior to selecting a colony. For slow-growing strains (small colonies after 48 hours) the following procedure is recommended: Inoculate 1 ml of Schaedler broth with one colony and incubate at 35–37°C for 5 hours. Flood a petri dish of Columbia Sheep blood agar with all this culture. Remove any excess liquid, and then incubate anaerobically for 18 hours at 35–37°C.

3. Place an API Rapid STREP test strip into an incubation tray.

4. With a pipet transfer approximately 0.1 ml (3 drops) of the bacterial suspension to fill the tubes and lower half of the cupules of the left half (VP-LAP) of the test strip. Completely fill only the tube of ADH. Avoid the formation of bubbles at the base of the tube by placing the pipet tip against the side of the cupule while tilting the strip slightly forward.

5. With pressure from your thumb, snap open a vial of the Rapid STREP medium. Take 0.5 ml of the bacterial suspension and add it to an ampule of Rapid STREP medium. Mix well.

6. With a pipet, transfer the Rapid STREP medium and suspension into the tube portions only of the right-hand side of the strip (*RIB* through *GLYG*).

7. Add mineral oil to the cupules of tests *ADH* to *GLYG*.

8. Place the cover over the incubation tray.

9. Incubate for 4 hours at 35–37°C to obtain the first reading and for 24 hours to obtain a second reading if this is required.

10. After 4 hours incubation, add the reagents to those cupules that require them according to the instructions in Table 70-3.

11. Wait 10 minutes.

12. Read the reactions as indicated in Table 70-3 and Color Plate 27. If necessary, expose the strip to a strong light (1000 W) for 10 seconds to decolorize any excess reagent in tubes *PYRA* through *LAP*.

13. Reincubation is necessary when the profile cannot be found in the index or when there is risk of confusion. Add water to the tray before reincubation.

14. After 24 hours of incubation, read the reactions *ESC*, *ADH*, and *RIB* through *GLYG*.

15. Note the results on the sheet provided.

16. If an identification code book is not available, you can compare your result with those given in Table 70-4.

Table 70-3 Interpretation of Results on the Rapid STREP Strip

Test	Substrate	Enzymatic Activity or Reaction Tested	Reagent To Be Added After 4 Hours of Incubation	Reaction (1) Positive	Reaction (1) Negative
VP	Pyruvate	Acetoin production	1 drop of VP 1 reagent 1 drop of VP 2 reagent wait 10 minutes	Pink-red	Colorless
HIP	Hippurate	Hydrolysis	2 drops of Ninhydrin reagent wait up to 10 minutes	Dark blue or violet	Colorless or pale blue
ESC	Esculin	β-Glucosidase		4h: Black or gray 24h: Black	4h: Colorless or pale yellow 24h: Colorless or pale yellow gray
PYRA	Pyrrolidonyl-2-naphthylamide	Pyrrolidonylarylamidase		Orange	Colorless or very pale orange
αGAL	6-Bromo-2-naphthyl-α-D-galactopyranoside	α-Galactosidase	1 drop ZYME A reagent + 1 drop of ZYME B reagent	Violet	Colorless
βGUR	Naphthol-AS-Bl-β-D-glucuronate	β-Glucuronidase		Blue	Colorless

Code	Substrate	Reaction	Positive / reading column		Negative / reading column	
			4h	24h	4h	24h
βGAL	2-Naphthyl-β-D-galactopyranoside	β-Galactosidase	Violet		Colorless or very pale violet	
PAL	2-Naphthyl phosphate	Alkaline phosphatase	Violet		Colorless or very pale violet	
LAP	L-Leucyl-2-naphthyl amide	Leucine arylamidase	Orange		Colorless	
ADH	Arginine	Arginine dehydrolase	Red-orange		Yellow	
RIB ARA MAN SOR LAC TRE INU RAF AMD	Ribose L-Arabinose Mannitol Sorbitol Lactose Trehalose Inulin Raffinose Starch (2)	Acidification	Orange-yellow (2)	Yellow (2)	Red	Orange-Red
GLYG	Glycogen		bright yellow		red or orange	

(For βGAL, PAL, LAP, ADH reactions: Wait 10 minutes (if necessary, decolorize with a bright light))

(1) During a second reading, after 24 hours of incubation, a deposit may be noticed in the tubes where the reagents ZYME A and ZYME B have been added. This phenomenon is normal and should not be taken into consideration.

(2) The acidification of starch is frequently weaker than that of other sugars.

Table 70-4 Differential Table: Positive Reactions Obtained with *Streptococci* After 4 to 24 Hours of Incubation at 35 – 37°C (%)

api Rapid STREP* V 2.0

	VP	HIP	ESC	PYRA	αGAL	βGUR	βGAL	PAL	LAP	ADH	RIB	ARA	MAN	SOR	LAC	TRE	INU	RAF	AMD	GLYG	βHEM	Dextran/Levane
Aerococcus viridans 1	1	50	96	96	1	8	1	0	0	0	99	1	99	96	96	99	8	8	96	0	0	NT
Aerococcus viridans 2	7	85	42	92	14	35	1	0	0	0	42	1	28	1	81	81	1	7	21	0	0	NT
Aerococcus viridans 3	11	94	99	52	94	41	11	0	0	0	11	11	94	5	76	99	47	99	23	0	0	NT
Streptococcus salivarius	97	0	98	0	33	0	65	40	100	0	0	0	3	0	84	75	60	87	45	1	1	+ (D)
Streptococcus mitis	0	0	1	0	1	0	47	61	100	13	6	0	0	0	95	18	10	6	61	10	0	
Streptococcus mutans	100	0	99	0	64	0	1	1	100	18	0	0	100	90	90	100	81	81	1	0	1	+ (L)
Streptococcus sanguis I/ 1	0	0	75	0	23	0	13	47	100	80	3	0	0	1	75	82	44	23	41	1	0	+ (D)
Streptococcus sanguis I/ 2	0	0	64	0	96	0	1	7	100	96	3	0	1	50	99	83	64	92	92	3	0	+ (D)
Streptococcus sanguis II	0	0	4	0	94	0	35	80	100	20	6	0	1	1	93	30	6	99	84	2	0	−
Streptococcus suis I	0	0	99	10	75	99	33	1	100	99	0	0	1	0	98	99	98	0	100	99	1	−
Streptococcus suis II	0	0	71	8	96	99	14	1	100	99	0	0	1	0	100	99	85	100	100	100	1	−
Streptococcus pneumoniae 1	0	0	18	50	97	0	70	1	100	50	1	0	1	0	100	95	68	99	90	40	0	−
Streptococcus pneumoniae 2	0	0	20	1	40	0	1	1	100	1	1	0	1	0	100	99	20	100	4	1	0	−
Streptococcus pyogenes 1	0	1	3	98	0	26	0	100	100	99	0	0	27	1	99	98	0	1	90	66	99	−
Streptococcus pyogenes 2	0	1	3	98	1	1	0	100	100	98	0	0	27	1	81	97	0	1	9	1	98	−
Streptococcus agalactiae	100	100	1	1	4	79	1	99	100	100	99	0	1	1	50	87	0	0	35	4	75	−
Streptococcus milleri I	100	0	1	0	0	0	7	99	100	99	0	0	1	1	1	85	0	1	2	0	90	−
Streptococcus milleri II	100	0	97	0	3	0	5	94	100	99	0	0	1	1	94	96	8	2	75	0	50	−
Streptococcus milleri III	100	0	99	0	97	0	5	99	100	100	0	0	99	1	96	96	6	96	99	0	40	−
Streptococcus gr. L	0	80	1	0	0	100	1	100	100	100	100	0	0	0	90	100	0	0	100	100	100	−
Streptococcus morbillorum	3	0	0	7	0	0	14	35	100	4	7	0	0	0	0	3	1	1	14	7	0	−
Streptococcus uberis	100	99	100	16	10	92	1	30	100	100	99	0	99	99	100	100	99	7	66	23	0	−
Streptococcus acidominimus	0	100	4	13	0	66	17	86	100	17	17	0	42	3	91	87	0	0	6	0	0	−
Streptococcus lactis/ diacetylactis	90	54	100	9	3	0	15	6	98	100	100	15	39	0	72	96	0	5	96	3	0	−
Streptococcus cremoris/ thermophilus	100	28	28	0	7	0	50	0	100	0	14	0	3	0	99	11	0	5	3	0	0	−
Streptococcus gr. E.P. & U	100	5	99	1	19	99	1	100	100	100	98	0	95	95	83	99	0	0	50	0	100	−
Streptococcus gr. G	0	8	12	0	15	68	6	100	100	100	97	0	0	0	55	85	0	2	96	13	100	−
Streptococcus equisimilis	0	1	9	0	0	100	1	100	100	97	97	0	0	0	45	100	0	1	96	10	97	−
Streptococcus zooepidemicus	0	0	15	0	0	100	1	99	100	100	85	0	0	99	100	0	0	0	99	99	100	−
Streptococcus dysgalactiae	0	0	0	0	0	100	0	100	100	100	100	0	0	50	96	100	0	0	99	30	0	−
Streptococcus equi	0	0	1	0	0	100	0	100	100	100	0	0	0	0	0	0	0	0	100	100	100	−
Streptococcus equinus	100	0	97	0	1	0	1	1	100	0	0	0	25	0	0	1	25	8	13	8	0	−
Streptococcus bovis I	99	0	100	0	76	2	1	0	97	0	0	14	97	0	100	100	83	99	100	97	1	+ (D)
Streptococcus bovis II/ 1	100	0	93	0	93	1	6	0	100	0	0	10	1	0	96	26	56	99	90	86	2	−
Streptococcus bovis II/ 2	100	3	100	0	78	41	38	0	100	0	0	1	3	0	99	79	32	79	11	1	1	−
Enterococcus faecalis 1	100	99	99	97	0	0	1	1	100	99	100	1	99	97	93	100	0	0	94	1	2	−
Enterococcus faecalis 2	100	1	100	97	0	0	3	1	100	99	100	1	99	97	93	100	0	0	94	1	2	−
Enterococcus faecalis 3	100	99	100	97	0	0	99	1	100	99	100	1	99	97	93	100	0	0	94	1	2	−
Enterococcus faecium 1	100	66	100	100	5	1	88	1	99	99	99	98	99	6	99	99	0	1	7	1	0	−
Enterococcus faecium 2	99	96	100	100	90	0	99	1	99	99	99	60	99	15	99	99	1	99	80	1	0	−
Enterococcus faecium 3	100	1	100	100	90	0	99	1	99	66	99	99	99	15	99	99	99	99	99	1	0	−
Enterococcus durans 1	100	25	100	100	1	0	99	1	99	100	100	8	1	0	100	50	0	1	30	0	96	−
Enterococcus durans 2	100	25	100	100	50	0	50	1	99	100	100	3	1	0	100	99	0	1	7	0	1	−
Enterococcus avium	100	60	100	95	6	0	1	1	100	0	100	40	100	99	100	100	1	40	11	0	0	−
Enterococcus gallinarum	100	100	100	100	100	83	100	0	100	100	100	100	100	0	100	100	100	100	83	16	0	−
Gemella haemolysans	54	0	0	36	0	0	0	45	33	9	9	0	99	45	36	9	0	0	9	0	0	−
Gardnerella vaginalis	0	99	0	0	0	1	53	0	99	0	46	6	1	0	1	0	0	0	73	53	0	NT
Listeria monocytogenes	100	97	100	0	0	0	0	0	97	0	2	0	0	0	42	100	0	0	88	0	5	NT

* 24-hr incubation profile; + (D) = production of dextran; + (L) = production of levan; and NT = not tested.
Explanation of symbols: ☐ 0 to 19% of positive answers; ▨ 20 to 79% of positive answers; and ▦ 80 to 100% of positive answers.

Rapid STREP Identification

If a codebook is available, use it. To use the codebook, one has to obtain a 7-digit numerical profile of the organism, as follows:

1. The 21 tests are divided into seven groups of three (a β-hemolysis result is added to the 20 tests of the strip).
2. A numerical value is given to each test. Negative tests are given the value of 0. Positive tests are given values according to the position of the test in its group:

   1 if the test is first in the group of three

   2 if the test is second in the group, and
   4 if the test is third in the group.

3. The values are totaled for each group of three. A 7-digit code is then obtained by listing these totals in sequence. This is the *numerical profile* of the organism, and it can be compared with the 7-digit code in the profile list in the codebook. *Note:* Each profile number is unique and represents only one sequence of reactions (and vice versa).

An example of a coded Rapid STREP form is given below.

* Rapid E, Rapid STREP, Rapid CH, STAPH Trac, Rapid SYS and Rapid NFT are trademarks of API System, S.A.

Normal Mouth, Nose, and
Throat Flora

Name _____

Desk No. _____

ÿapi®
STAPH-IDENT™
System

| REF. NUMBER | PATIENT | SEX/AGE | SOURCE/SITE |
| DATE RECEIVED | DEPT./SERVICE | | PHYSICIAN |

	PHS 1	URE 2	GLS 4	MNE 1	MAN 2	TRE 4	SAL 1	GLC 2	ARG 4	NGP 1
RESULTS										

PROFILE NUMBER

GRAM STAIN ☐   COAGULASE ☐   Additional Information   Identification

MORPHOLOGY ☐   CATALASE ☐

42-100 (11/85)

ÿapi®Rapid STREP*

Ref. _____ Patient _____
Date _____ Source _____
Dr. _____ Service _____

	VP	HIP	ESC	PYRA	α GAL	β GUR	β GAL	PAL	LAP	ADH	RIB	ARA	MAN	SOR	LAC	TRE	INU	RAF	AMD	GLYG	β HEMOL
4 H																					
24 H																					

Gram ☐   Hemolysis ☐   Identification

Morphology ☐   Dextran-Levan ☐

Catalase ☐   Other Tests: ☐

Serology

Printed in France
480004 A 05/85

***** Rapid E, Rapid STREP, Rapid CH, STAPH Trac, Rapid SYS and Rapid NFT are trademarks of API System, S.A.**

Draw a picture of the different types of colonies of bacteria found in your mouth, nose, and throat isolates on blood agar.

Draw the cell morphology and grouping of bacteria that you observed in gram-strain preparations.

Look up the description of two of your isolates in a standard reference such as *Bergey's Manual of Systematic Bacteriology*. Summarize your findings below and discuss your results.

71. *Sensitivity Discs in the Therapeutic Use of Antibiotics: Kirby–Bauer Technique*

The sensitivity-disc method of determining which antibiotic is effective against an organism is a rapid, accurate, and inexpensive diagnostic tool. Broad spectrum antibiotics affect a wide variety of both gram-negative and gram-positive organisms, but other antibiotics inhibit fewer organisms. Antibiotic mechanisms vary from acting as structural analogues to inhibiting protein synthesis (streptomycin).

Filter-paper discs impregnated with various antibiotics are used to determine the in vitro susceptibility of clinical isolates to antibiotics. Information about the sensitivity of a pathogenic organism can help the physician choose antibiotics for treating the patient from whom the specimen was isolated.

Strains are increasingly isolated that are resistant to several antibiotics. Multiple resistance is conferred by genes carried on plasmids passed from one strain to another by resistance transfer factors. Thus, in addition to using information about the sensitivity of a clinical isolate to certain antibiotics, the physician can find the pattern of drug resistance of an isolate, the **antibiogram**, a useful epidemiological tool in tracing an outbreak of an organism.

To examine the susceptibility of an organism to antibiotics, discs impregnated with various antibiotics are placed on the surface of an agar plate seeded with the organism being examined. The organism grows during incubation, but zones of no growth develop around the discs containing antibiotics that inhibit growth. The size of the zone of inhibition caused by the diffusion of the agent into the agar is directly related to the degree of susceptibility of the organism.

Inhibition zone size is affected by technical variables, including inoculation size, incubation time and temperature, medium composition, pH, gaseous atmosphere, stability of the antibiotic, and others. The use of the carefully standardized techniques reduces the effects of variables.

It is important that a number of colonies from the culture of interest be sampled to avoid the chance selection of variants that have lost resistance, but the final inoculum must be heavy in order to increase the possibility of detecting a mutation to resistance. Designated American Type Culture strains must be used routinely in the procedure to assure that all the variables in the procedure are controlled so that results are reproducible.

This exercise illustrates the differences in sensitivity of gram-positive and gram-negative bacteria to several antibiotics. You will measure the antibiotic sensitivity of only one bacterium. Students at odd-numbered desks will test a gram-negative species, and those at even-numbered desks will test a gram-positive species.

PROCEDURE

1. Dip the sterile swab into the bacterial broth culture of standard density; then rotate it several times, with firm pressure on the inside wall of the tube above the fluid level, to remove excess inoculum.
2. Streak the swab over the entire Mueller–Hinton agar plate surface and repeat two more times, rotating the plate about 60 degrees to insure even distribution of inoculum. Replace plate top and allow 3–15 minutes for excess surface moisture to absorb.
3. Apply the antibiotic discs to the surface with sterile forceps or dispenser. Distribute the discs evenly. Gently press down each disc with the forceps to insure complete contact with the agar surface (Figure 71-1).
4. Within 15 minutes after the discs are applied, incubate the plates at 37°C.
5. After 16–18 hours of incubation, examine each plate and measure the diameters of the zones of complete inhibition using a ruler or template (Figure 71-2).
6. Interpret the zones of inhibition by referring to Table 71-1 provided and report the organisms that tested as susceptible, intermediate, or resistant to the various agents used (Figure 71-3).
7. Obtain the plate of the second bacterium from another student and measure the zones

Figure 71-1 Application of antibiotic discs to test culture individually and by dispenser.

Figure 71-2 Zones of growth inhibition due to antibiotics.

Zone size
interpretation

R = Resistant; Zone of inhibition with a diameter equal to or less than inner circle.

I = Intermediate; Zone of inhibition with a diameter greater than R, but less than outer circle.

S = Sensitive, Zone of inhibition with a diameter greater than outer circle.

Figure 71-3 Interpretation of zone of inhibition caused by antibiotics.

of inhibition to determine the susceptibility of this organism to the various antibiotics.

Quality-Control Procedures

ATCC cultures *S. aureus*, *E. coli*, and *P. aeruginosa* will be used by your instructor, employing the same materials and procedures as you used. Results of the control test will be posted for use in interpreting your test data.

QUESTIONS

1. Why do some colonies grow inside the zone of growth inhibition?

2. Are all antibiotics equally effective against both gram-positive and gram-negative organisms?

3. How could one determine if the agent is inhibiting or killing the organisms?

4. Why don't in vitro results guarantee clinical success?

5. Will all strains of an organism react in the same manner to an array of antibiotics? Why or why not?

6. How are antibiotics used other than for treatment of disease?

Table 71-1   Interpretation of Zone Diameters of Test Cultures

	Abbr	*Resistant (mm or less)*	*Intermediate (mm range)*	*Susceptible (mm or more)*
Ampicillin	AM	20	21–28	29
Amikacin	AN	14	15–16	17
Aureomycin	A	14	15–18	19
Bacitracin	B	8	9–12	13
Cephalothin	CR	14	15–17	18
Chloramycetin	C	12	13–17	18
Colymycin	CL (colistin sulfate)	8	9–10	11
Erythromycin	E	13	14–17	18
Gantrisin	G	12	13–16	17
Gentamicin	Gm	12	13–14	15
Kanamycin	K	13	14–17	18
Lincomycin	L	13	14–17	18
Methicillin	ME	9	10–13	14
Nalidixic Acid	NA	14	14–18	19
Novobiocin	NB	17	18–21	22
Oxacillin	OX	9	10–13	14
Penicillin	P	20	21–28	29
Polymyxin B	PB	8	9–11	12
Streptomycin	S	11	12–14	15
Sulfadiazine	SD	12	13–16	17
Triple sulfa	SSS	10	11–15	16
Tetracycline	TE	14	15–18	19
Tobramycin	TM	12	13–14	15

Chloramphenoel
Neomycine

72. *Demonstration of Koch's Postulates**

Koch's postulates are a cornerstone of medical microbiology. These postulates, which make it possible to establish that a certain microorganism is the causative agent of a given disease, are as follows:

1. The microorganism is regularly found in association with the specific disease.
2. The microorganism can be isolated in pure culture on laboratory media.
3. Inoculation of this pure culture into experimental animals produces a disease similar to the original one.
4. The microorganism can be isolated from the animals that have the induced disease.

In this exercise you will have the opportunity to isolate a microorganism from an infected goldfish, to apply Koch's postulates, and to demonstrate that this organism is the cause of the disease. In addition, you will determine the antibiotic sensitivity of the isolated pathogen as part of its characterization and demonstrate the ability of an effective antibiotic to protect a goldfish against this pathogen.

PROCEDURE

Students will work in groups of three or more.

First Laboratory Period

1. Each group of students is given a dead goldfish that expired from a *Vibrio anguillarium* infection. Place the dead fish on a paper towel saturated with 1% phenol or other disinfectant. With a small scissors, sterilized in ethanol or boiling water, slit the belly of the fish from its anus to its gill. From this slit make vertical cuts to the spine just behind the gill and just at the beginning of the tail region (Figure 72-1A). Do not rupture the gut! With a sterile forceps, open the dissected fish to expose the internal organs.

---

*This exercise has been adapted from "Infection of Goldfish with *Vibrio anguillarium*" by W. W. Umbreit and E. J. Ordal. *ASM News* 38 (Feb. 1972):93–98. Permission for its use was granted by The American Society for Microbiology.

2. Insert a sterile swab into the peritoneal cavity and then use the swab to streak a 1% brain–heart infusion agar plate containing 1.5% salt and 0.5% starch (Figure 72-1B). One plate should be streaked heavily in two to five directions. On the surface of this latter plate, place a series of antibiotic discs (Exercise 71) to determine the antibiotic sensitivity of the gut microflora. Using your inoculating loop, streak a second brain–heart infusion salt–starch plate to obtain isolated colonies. Incubate the agar plate at 30°C for 24–48 hours. Also prepare and observe a gram stain of the microflora of the peritoneal fluid.

Second Laboratory Period (Two days later, during the week between labs at some institutions)

3. After incubation of the agar plates, observe and record the antibiotic sensitivity of the isolate.
4. Prepare and observe a gram stain and run oxidase tests on several isolated colonies. From a colony of oxidase-positive, gram-negative curved rods, inoculate one screw-cap tube of brain–heart infusion broth containing 1.5% NaCl. Finally, flood the plate with Gram's iodine to determine if the colony is capable of hydrolyzing starch.
5. Incubate the tube of brain–heart infusion broth on a tube shaker for 48 hours at 30°C.

Third Laboratory Period

6. Each student group is provided with three goldfish, two of which are in a beaker of water and the third fish is in a beaker containing 100 μg/ml of tetracycline. With a sterile 1 ml hypodermic syringe, inject 0.1 ml aliquots of the brain–heart infusion broth culture into the peritoneal cavities of one of the fish from the beaker of water and the fish from the antibiotic solution. Inject the second fish from the water beaker with 0.1 ml of sterile saline. For these injections, the fish are handled in a moist towel to pre-

Figure 72-1   Koch's postulate experiment. *A*: Dissecting the dead goldfish. *B*: Swabbing exudate from the peritoneal cavity. *C*: Injecting a goldfish with the culture.

vent injury. Hold each fish on its back and inject just in front of or just behind the ventral fin (Figure 72-1C). Return the fish to their respective beakers. *Caution*: The injection should be almost parallel to the body. Do not injure the fish internally by injecting it too deeply or at a right angle. Do not keep the fish out of the water for more than 5 minutes. Incubate the fish in their beakers at room temperature for 24–48 hours.

Fourth Laboratory Period (24–48 hours later, during the week at some institutions)

7. After incubation, remove and dissect any dead fish (or refrigerate until dissection is scheduled). Streak the peritoneal exudate on brain–heart infusion salt-starch plates and determine the morphology, gram reaction, oxidase and starch reactions, and antibiotic sensitivities of the isolates as outlined in Steps 2–4.

Fifth Laboratory Period 3

8. Compare the characteristics of the isolate taken from the original dead fish with the characteristics of the organisms recovered from the fish you injected.
9. Record your findings on the report sheet.

QUESTIONS

1. Why is it important to culture the infectious organism and use it to reinfect the host?
2. How might you demonstrate whether or not a toxin caused the death of these fish?
3. Why must you find the organism in every case of the disease?

Demonstration of Koch's Postulates

Name _____

Desk No. _____

Which fish died?

What caused the death of these fish?

In this table, compare the characteristics of the isolate taken from the original dead fish with the characteristics of the organisms recovered from the fish you injected.

Characteristics	Original isolate	Final isolate
Morphology		
Gram reaction		
Oxidase reaction		
Starch hydrolysis		
Antibiotic sensitivity*		
Penicillin		
Novobiocin		
Tetracycline		
Streptomycin		

*Record diameter of zone of inhibition.

73. *Complement Fixation Test*

The complement system is a group of 11 related proteins normally found in the serum of higher vertebrates. The system can be activated by two pathways and is involved in many immunological reactions. One important reaction leads to the lysis of cell membranes to which the complement complex has been bound.

This complement fixation reaction is a sensitive method of detecting the presence of antibodies in serum. The visualization at the end of the test is very simple and is not susceptible to misinterpretation. For this reason alone the complement fixation test is widely used in clinical laboratories. The complement fixation reaction involves two separate antigen–antibody reactions: the test system and the indicator system as shown in Figure 73-1.

The test system consists of (1) serum that may or may not contain antibodies against the antigen used in the test, (2) the antigen, and (3) a known amount of complement. Because individuals may have quantitative differences in their own complement levels, the test serum (1) is first heated gently to inactivate some of its complement proteins. Then a

known amount of complement is added. If the antibody reacts with the antigen, then all the complement is bound in the reaction: It is used up. How do we know if it is used up or not? That is the purpose of the second, or indicator, system.

The indicator system has two parts: (1) sheep red blood cells and (2) antibody against sheep red blood cells that has been made in a rabbit or goat. If complement is added to the mixture of (1) and (2), the attack complex formed by the complement system will lyse the red blood cells. It is called an indicator system because it tells us whether the complement was used up in the test system. If it was not, the cells will lyse, indicating that the antibody and antigen were not matched in the test system.

Procedure

1. Take six 2 ml capacity test tubes and place them in a rack. Label them as follows:
   1. Complete test system, Patient A

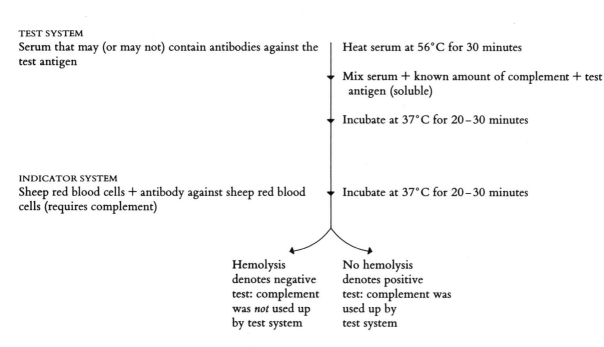

TEST SYSTEM
Serum that may (or may not) contain antibodies against the test antigen

Heat serum at 56°C for 30 minutes

Mix serum + known amount of complement + test antigen (soluble)

Incubate at 37°C for 20–30 minutes

INDICATOR SYSTEM
Sheep red blood cells + antibody against sheep red blood cells (requires complement)

Incubate at 37°C for 20–30 minutes

Hemolysis denotes negative test: complement was *not* used up by test system

No hemolysis denotes positive test: complement was used up by test system

Figure 73-1   The complement fixation reaction

Table 73-1 Complement Fixation Test System Reagents

Reagent	Tube number					
	1	2	3	4	5	6
TEST SYSTEM						
Kolmer saline (ml)	—	—	—	—	0.3	0.2
Antigen (ml)	0.1	0.1	0.1	0.1	—	—
Complement (1:30) (ml)	0.1	0.1	0.1	0.1	—	0.1
Antibody (ml)	PAT.	PAT.	POS.	NEG.		
	A	B	CONTROLS		—	—
	0.1	0.1	0.1	0.1	—	—
INDICATOR SYSTEM						
2% sheep red cells (ml)	0.1	0.1	0.1	0.1	0.1	0.1
Hemolysin (anti-sheep RBC antibody) (ml)	0.1	0.1	0.1	0.1	0.1	0.1

2. Complete test system, Patient B
3. Positive control
4. Negative control
5. Red cell control
6. Complement control

2. Pipet 0.1, 0.2, or 0.3 ml of the test system reagents (Kolmer saline, antigen, complement 1:30, and antibody) in the quality indicated in Table 73-1. (Do not add the indicator system at this time.)
3. Mix the reagents by inversion or with the aid of a vortex mixer.
4. Incubate the tubes for 20–30 minutes in a 37°C water bath
5. Add 0.1 ml of the two reagents for the indicator system to all six tubes.
6. Mix by inversion or vortex and continue incubation at 37°C for an additional 20–30 minutes.
7. After incubation, examine the tubes to see if the contents are still opaque (no lysis) or translucent (hemolysis). Record your results and interpret them.

QUESTIONS

1. Why is it necessary to have so many control tubes in the experiment? What are the purposes of each one?
2. Why is the complement fixation test such a popular method in the diagnostic laboratory?

74. *Detection of Chemical Carcinogens: Ames Test*

The elimination or cancer-causing agents from our environment is the serious concern of many scientists. A large proportion of chemical carcinogens have been shown to be mutagens. One plausible explanation of how carcinogens act is that they damage and cause mutations in mammalian DNA. From cost and animal-welfare viewpoints, it is far easier and faster to screen for mutagenicity in microbial systems than it is in rats, mice, or other animal models.

A highly reliable microbial test for detecting potential carcinogens is the *Salmonella*/mammalian-microsome test developed by Dr. Bruce Ames and colleagues. The Ames test for detecting mutagenic agents, and hence probable carcinogens, employs special histidine-dependent strains of *Salmonella typhimurium.** When these histidine-requiring cells are inoculated on a medium containing a trace of histidine, only a few cells revert to histidine-independence and are able to form colonies. If a mutagenic chemical is added to this histidine-deficient medium, the reversion rate to histidine-independence is greatly increased. Although not sufficient for colony formation, the trace levels of histidine in the test medium enable all cells to undergo a few cell divisions, which in some cases is necessary for mutagenesis. In our experiment, we will use a leucine-independence reverting strain, isolated by another researcher, Dr. Paul Margolin.

Many carcinogenic chemicals are not carcinogenic (or mutagenic) unless they are metabolized to active substances. In people and animals these metabolic conversions to active carcinogens occur primarily in the liver. Thus, to enhance the conversion of potentially mutagenic or carcinogenic chemicals to their active forms, agents in the Ames test are usually treated with a rat-liver homogenate prior to their addition to the culture medium. The particular class of carcinogens used in this exercise, namely the nitrocarcinogens, are converted to their active forms by nitroreductases of the bacteria themselves, thus dispensing with the need for use of a rat-liver homogenate in this exercise.

PROCEDURE

1. With a sterile glass spreader, plate 0.1 ml suspensions of the leu-130 *Salmonella typhimurium* ATCC 49416 provided on three plates of SEM (semienriched minimal) agar.
2. Aseptically apply a sterile filter disc to the center of the lawn on the plates.
3. Label one plate *control*.
   Label one of the other plates *MMS* (methane sulfonic acid methyl ester) and the other *EMS* (methane sulfonic acid ethyl ester).
4. **Put on protective gloves!**
5. With a sterile dropper, add 1 or 2 drops of sterile saline to the control disc and add 1 or 2 drops of MMS to the second plate. With a fresh sterile pipet, add EMS to the disc on the third plate. Carefully place both pipets and your gloves in special plastic bags provided for hazardous materials.
6. Incubate the agar plates at 37°C for 48 hours.
7. Observe the agar plates for large colonies around the discs, demonstrating back-mutation revertants.
8. Record the numbers of large colonies around each disc and on the minimal agar plate not containing any chemical agents.
9. Controls using nutrient agar will be prepared by your instructor. Observe these plates. Are there differences in growth on the nutrient agar and minimal agar without added chemicals? Do any of the chemical agents tested cause growth inhibition on the nutrient agar?
10. Record your findings on the report sheet.

---

*These mutant strains made available by Dr. Bruce Ames have point mutations, that is, deletions, in their histidine operon because such deletions do not commonly revert. These mutants also lack a DNA excision-repair mechanism so DNA errors are not corrected, thus enhancing the strains' sensitivity to mutagens. Furthermore, these mutants have a defective lipopolysaccharide layer on their cell surface that enable mutagens to more easily penetrate into the cell. Some strains used in this test also carry R plasmids, which make these strains more sensitive to some weak mutagens.

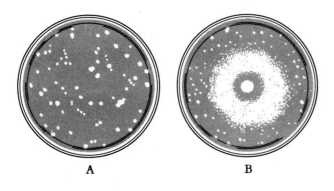

Figure 74-1 The Ames spot test to demonstrate mutagenicity. *A*: Control showing spontaneous revertants. *B*: Revertants caused by mutagen in center.

QUESTIONS

1. In the Ames test, mammalian-liver homogenates are usually employed. Why? Why weren't they used in this exercise?
2. Explain the importance of controls in this exercise.
3. If you inspect the plates carefully, you will see a slight haze of confluent background growth that is readily distinguished from the large revertant colonies. What is the explanation for the haze?

Detection of Chemical Carcinogens: Ames Test

Record the number of large colonies around each chemical-containing disc and on the agar plates not containing any chemical agents.

Chemical agent	Minimal agar	Nutrient agar
None, saline		
MMS		
EMS		

Explain the advantages and disadvantages of using the Ames test for detecting carcinogens.

75. *Snyder Caries Susceptibility Test*

Caries (tooth decay) is a bacterial disease in which various species of *Streptococcus* demineralize tooth surfaces by their catabolism of carbohydrates to acidic products (lactic, formic, acetic, and butyric acids). Dental plaque is composed mainly of bacteria and their products. In addition to *Streptococcus mitis, S. salivarius, S. sanguis,* and *S. mutans,* species of *Actinomyces, Fusobacterium,* and *Veillonella* are commonly found in plaque. *Streptococcus mutans* seems the most important in the disease process. Characteristically plaque streptococci produce capsular polysaccharides composed of soluble and insoluble dextranase-sensitive polymers of glucose and levan polymers of fructose which cause the bacteria to adhere to the teeth and form plaque.

The pH at which demineralization of teeth takes place is approximately 5.6. In this exercise we will use the Snyder Fermentation test which is based on the capacity of bacteria from the mouth to catabolize glucose to produce the quantity of acids necessary to demineralize tooth enamel.

PROCEDURE

1. Aseptically, with an alcohol sterilized forcepts take a small (1 cm³) sterile cube of paraffin wax from its container. Place the cube in your mouth and chew it for three minutes. As you are chewing it, move it around in your mouth so that it makes contact with the surfaces of many teeth. Take care not to swallow your saliva.
2. Collect your saliva in a beaker.
3. Inoculate 0.2 ml of your saliva into a tube of sterile liquified Snyder test medium which will be kept in a water bath at 50°C until you use it.

4. Vigorously mix the inoculum into the medium by rolling the tube between the palms of your hands. This should mix the inoculum in the medium without entraining air bubbles in the medium.
5. After the medium solidifies place the tube in a 37°C incubator.
6. Examine the tube for color change after 24, 48, and 72 hrs. Snyder test medium has bromcresol green as a pH indicator which changes from green (pH 4.8) to yellow (pH 4.4 or lower) as acid is produced by the bacteria. The time taken for the medium to change is used as an indication of caries susceptibility as follows: (a) change in 24 hours—marked susceptibility; (b) change in 48 hours—moderate susceptibility; (c) change in 72 hours—slight susceptibility; (d) no change in 72 hours—negative. Record your results.

QUESTIONS

1. The hamster is a laboratory animal that is susceptible to dental caries. Design an experiment to demonstrate that it is *Streptococcus mutans* which causes dental caries.
2. As an addition to the experiment in question 1, design an experiment to show that dietary sugars can be a factor in caries formation.
3. Why does regular toothbrushing reduce dental plaque?
4. Do saccharin and aspartame have the same potential as sucrose for promoting dental plaque?

A P P E N D I X

Stains and Reagents

Acid alcohol (Ex 14)
3 ml HCl
97 ml 95% ethyl alcohol
Add the HCl to the alcohol.

Acridine orange (Ex.5)
0.1 g acridine orange
10 ml distilled water
Store in a foil-wrapped tube.

Acriflavin (Ex. 50)
2 mg acriflavin
20 ml distilled water
Filter, sterilize, and store in a tube wrapped in foil. Use gloves to make up and handle this solution.

Bouin's fixative (Ex. 62)
100 ml picric acid in buffered concentrated formalin
5 ml glacial acetic acid
Add acetic acid just before use.
Caution: Picric acid is explosive when dry.

Brom thymol blue (Ex. 25)
0.4 g brom thymol blue
500 ml 95% ethanol
500 ml distilled water
Dissolve the dye in alcohol and then dilute with the water.

Carbol fuschin (Ex. 12, 14)
0.3 g basic fuschin
10 ml 95% ethyl alcohol
5 ml phenol
95 ml distilled water
Dissolve the basic fuschin in alcohol. Heat the phenol crystals to 45°C to melt them. Add 5 ml phenol to the water and mix. Mix the solutions together and allow the mixture to stand for several days. Filter the reagent before use.

Congo red (Ex. 15)
0.5 g congo red
100 ml distilled water

Crystal violet (Ex. 12, 13)
2 g crystal violet
20 ml 95% ethyl alcohol
0.8 g ammonium oxalate
80 ml distilled water
Dissolve crystal violet in alcohol and ammonium oxalate in water. Allow ammonium oxalate solution to stand overnight or heat gently until in solution. Mix the two solutions together and filter.

Diphenylamine (Ex. 69)
0.7 g diphenylamine
60 ml concentrated sulfuric acid
28.8 ml distilled water
11.3 ml concentrated hydrochloric acid
Dissolve the diphenylamine in sulfuric acid, and then add the water. Cool the mixture slowly, add the hydrochloric acid and allow to stand overnight.

FA buffer (Ex. 61)
It is a 0.85% saline solution buffered with phosphate to pH 7.2. This buffer can be ordered from DIFCO, catalog #2314-15-0.

FA mounting fluid (Ex. 61)
glycerin adjusted to pH9
This reagent can also be ordered from DIFCO, catalog #3340-57-5.

Ferric chloride (Ex. 36)
10 g ferric chloride
100 ml water
Store in refrigerator.

FITC conjugate antiserum (Ex. 61)
FA *Salmonella* poly
This reagent can be ordered from DIFCO, catalog #3187-56.

Gram's iodine (Ex. 13)
1.0 g iodine crystals
2.0 g. potassium iodide
300 ml distilled water
3 g sodium bicarbonate

Hydrogen peroxide (3%) (Ex. 38, 43, 46)
3 ml hydrogen peroxide
97 ml distilled water

Table A-1 Nephelometer Standards

	Tube number									
	1	*2*	*3*	*4*	*5*	*6*	*7*	*8*	*9*	*10*
1% Barium chloride (ml)	0.1	0.2	0.3	0.4	0.5	0.6	0.7	0.8	0.9	1
1% Sulfuric acid (ml)	9.9	9.8	9.7	9.6	9.5	9.4	9.3	9.2	9.1	9
Approximate cell density ($\times 10^8$/ml)	3	6	9	12	15	18	21	24	27	30

Kirkpatrick fixative (Ex. 61)
  60 ml isopropyl alcohol
  30 ml chloroform
  10 ml formaldehyde (37%)

Kovacs reagent (Ex. 36)
  50 g p-dimethylaminobenzaldehyde
  750 ml amyl or butyl alcohol
  250 ml concentrated hydrochloric acid
Prepare this reagent in a well ventilated chemical hood.

Malachite green (Ex. 15, 67)
  5 g malachite green oxalate
  100 ml distilled water

Maneval's stain (modified) (Ex. 15)
  0.05 g fuschin
  3 g ferric chloride
  5 ml glacial acetic acid
  3.9 ml liquified phenol
  95 ml distilled water
Prepare this stain in a well ventilated chemical hood

McFarland nephleometer standards
  See Table A-1

Methylene blue stain (Ex. 12, 14)
  8.9 g methylene blue
  600 ml ethyl alcohol
  2.0 g KOH
  2000 ml distilled water
  Combine and filter.

Naphthol (Ex. 35)
  5 g α-naphthol
  100 ml 95% ethanol

Nessler's reagent (Ex. 69)
  A: 35 g potassium iodide
    50 g mercuric iodide
    200 ml distilled water
  B: 50 g sodium hydroxide
    250 ml distilled water

Dissolve solutions A and B and cool. Add B to A in 500-ml volumetric flask, shaking. Add distilled water to 500 ml. Allow precipitate to settle for 1 week before using. Decant the clear solution and keep in well stoppered bottles away from light.

Nitrate reagents (Ex. 69)
  A: Sulfanilic acid
    8 g sulfanilic acid
    1000 ml 5N acetic acid
  Store in glass-stoppered bottle.
  B: α-naphthylamine
    5 g α-naphthylamine (may be carcinogenic)
    1000 ml 5N acetic acid
  Heat gently to dissolve. Filter and store in a brown, glass-stoppered bottle. Refrigerate.

p-Nitrophenylphosphate reagent (Ex. 48)
  2 g p-nitrophenylphosphate
  100 ml distilled water

Oxidase reagent (Ex. 38)
  0.1 g dimethyl- (or tetramethyl-) para-phenylenediamine hydrochloride
  10 ml distilled water
  The tetramethyl reagent is less toxic and less stable than the dimethyl reagent. Either reagent should be made fresh and stored in the refrigerator for no longer than 1 week.

Potassium hydroxide creatine (Ex. 35)
  40 g KOH
  0.3 g creatine
  100 ml water
  Store in refrigerator and use within 1 week.

Safranin (Ex. 12, 15)
  2.5 g safranin-0
  100 ml ethanol
  900 ml distilled water

Saline (Ex. 3, 26)
  9 g NaCl
  100 ml distilled water

Trommsdorf's solution (Ex. 69)

20 g $ZnCl_2$

4 g starch

2 g zinc iodide or potassium iodide

Dissolve $ZnCl_2$ in 100 ml distilled water. Dissolve starch in 150 ml distilled water. Heat zinc mixture to boiling, and add to starch mixture while stirring and heating until solution is nearly clear. Add 300 ml distilled water and the zinc or potassium iodide. Dilute to 1 liter, filter, and store in well stoppered bottles in the dark.

Vaspar (Ex. 35, 36)

1 lb vaseline

1 lb paraffin

Melt and combine the vaseline and pariffin. Pour into tubes with the aid of a small funnel.

Media

Ames test minimal medium (Ex. 74)

20 g glucose

0.3 g $MgSO_4 \cdot 7\ H_2O$

3 g citric acid

15 g $K_2\ HPO_4$

5.2 g $Na(NH_4)HPO_4 \cdot 4\ H_2O$

15 g agar

1000 ml distilled water

Autoclave for 15 minutes and dispense into petri plates.

Ames test — soft agar

Difco agar, 0.6 g; NaCl, 0.5 g; distilled water, 100 ml; autoclave for 10 minutes.

Ames test — biotin-histidine solution

Biotin, 0.0122 g; histidine-hydrochloride, 0.0077 g; distilled water, 100 ml; autoclave for 10 minutes.

Apple juice medium (Ex. 60)

0.4 g $(NH_4)_2HPO_4$

100 ml commercial apple juice with no preservatives

APT medium (Ex. 46)

7.5 g yeast extract

12.5 g tryptone

10 g dextroase

5 g sodium citrate

0.001 g thiamine hydrochloride

5 g NaCl

5 g K_2HPO_4

0.14 g $MgSO_4$

0.05 g ferrous sulfate

0.2 g sorbitol monooleate complex

1000 ml distilled water

Add 15 g agar per liter for *APT agar.*

Add 2 ml BCP indicator solution per liter of APT broth.

Add 0.3% sterile $CaCO_3$ to APT agar for *APT $CaCO_3$* and dispense while stirring.

Azide yeast extract broth (Ex. 46)

15 g polypeptone

4.5 g beef extract

7.5 g yeast extract

7.5 g NaCl

0.2 g sodium azide

1000 ml distilled water

BCP carbohydrate media (Ex. 28, 43, 45, 50, 60)

Basal broth: 10 g tryptone

5 g yeast extract

2 g K_2HPO_4

2 ml BCP indicator solution

1000 ml distilled water

BCP indicator solution: 0.4 g brom cresol purple

7.4 ml 0.1 N NaOH

9992.6 ml 50% ethanol

For *BCP agar* add 15 g agar per liter. Adjust the reaction to pH 7.1–7.2. Dispense in tubes with inverted insert tubes and sterilize at 121°C for 15 minutes (see exceptions noted below).

Glucose, lactose, sucrose, and mannitol are used in a final concentration of 1%. Other carbohydrates such as dulcitol, salicin, and so on are used in a final concentration of 0.5%. Glucose, mannitol, dulcitol, salicin, adonitol, and inositol may be added to the basal medium prior to sterilization. Medium containing neutral glycerol should be sterilized at 121°C for 10 minutes. Disaccharides such as lactose, sucrose, and cellobiose (10% solution in distilled water, neutral pH) should be sterilized by filtration or at 121°C for 10 minutes added to previously sterilized basal medium. Arabinose, xylose, and rhamnose also should be sterilized separately. Add one-tenth the volume of the sterile aqueous carbohydrate solution to the sterile basal medium. For example, if basal medium is tubed in 3.0-ml amounts, add 0.3 ml of sterile aqueous carbohydrate solution. Note: Raffinose must be filter sterilized.

Blood agar (Ex. 38)

10 g tryptone

3 g beef extract

5 g NaCl

15 g agar

1000 ml distilled water

Sterilize and cool to 45°C. Add 50 ml fresh defibrinated blood.

Brain-heart infusion media (Ex. 72)
37 g brain-heart infusion powder
1000 ml distilled water
For *BHI broth plus 1.5% NaCl,* add 15 g NaCl per liter.
For *BHI salt-starch agar*, add 15 g NaCl, 5 g starch, 15 g agar per liter.

Casein agar (Ex. 33, 36, 39, 43)
5 g tryptone
2.5 g yeast extract
1 g glucose
15 g agar
20 ml skim milk
980 ml distilled water
Add skim milk just before dispensing into tubes. Autoclave 10 minutes. Cool rapidly.

Corn meal agar (Ex. 67)
50 g corn meal
15 g agar
1000 ml distilled water
Heat to boiling to dissolve. 1% Tween 80 may be added to promote chlamydospore formation by *Candida albicans.*

Decarboxylase base broth (Ex. 36)
5 g thiotone peptone
5 g beef extract
0.10 g brom cresol purple
0.005 g cresol red
0.5 g dextrose
0.005 g pyrodoxal
1000 ml distilled water
Adjust pH to 6.0.

Deca-strength broth (Ex. 54)
100 g peptone
50 g yeast extract
25 g NaCl
80 g K_2HPO_4
1000 ml distilled water
Adjust pH to 7.6.

Desoxycholate agar (Ex. 59)
10 g peptone
10 g lactose
1 g sodium desoxycholate
5 g NaCl
2 g K_2HPO_4
1 g ferric citrate
1 g sodium citrate
0.03 g neutral red

15 g agar
1000 ml distilled water
Pour a thin layer of clear agar (1.5%) on top of hardened plates to eliminate any surface effects of colony growth.

DNase-test agar (Ex. 45, 70)
20 g bacto tryptose
2 g DNA
5 g NaCl
15 g agar
0.05 g methyl green
1000 ml distilled water
Boil the first five components in the distilled water and autoclave.

Egg yolk agar (Ex. 37, 44)
5 g tryptone
2.5 g yeast extract
1 g glucose
15 g agar
100 ml 5% egg yolk solution
900 ml distilled water

Emerson agar (Ex. 33)
4 g yeast extract
15 g soluble starch
1 g K_2HPO_4
0.5 g $MgSO_4 \cdot 7H_2O$
20 g agar
1000 ml distilled water
Emulsion agar is agar for a pour plate.

Endo agar (Ex. 57, 58)
10 g peptone
10 g lactose
3.5 g K_2HPO_4
2.5 g sodium sulfite
0.5 g basic fuschin
15 g agar
1000 ml distilled water

Eosin-methylene-blue agar (EMB) (Ex. 40, 41, 57)
10 g peptone
5 g lactose
5 g sucrose
2 g K_2HPO_4
0.4 g eosin Y
0.06 g methylene blue
15 g agar
1000 ml distilled water

Flo agar See *Pseudomonas agar F*

Glucose-acetate medium (Ex. 67)
1 g glucose
2.5 g yeast extract
8.2 g sodium acetate (or 2.3 ml glacial acetic acid)
15 g agar
1000 ml distilled water
Adjust pH to 4.8.

Glucose agar (Ex. 21, 22, 23)
Nutrient agar plus 5 g glucose per liter

Glucose broth (Ex. 20, 22)
Nutrient broth plus 5 g glucose per liter

Grape juice medium (Ex. 60)
4 g $(NH_4)_2HPO_4$
1000 ml commercial grape juice with no preservatives

Heart infusion agar (Ex. 52)
25 g heart infusion powder
15 g agar
1000 ml distilled water
Adjust pH to 7.0.

KF agar (Ex. 58)
10 g polypeptone peptone
10 g yeast extract
5 g NaCl
10 g glycerophosphate
0.636 g Na_2CO_3
20 g maltose
1 g lactose
0.4 g sodium azide
0.018 g phenol red
10 g agar
1000 ml distilled water
Adjust pH to 7.2.

Lactose broth (Ex. 57)
3 g beef extract
5 g peptone
5 g lactose
1000 ml distilled water
In water analysis, use double-strength lactose broth for large tubes to which 10 ml water samples are added.

Lactose yeast-extract agar (Ex. 39)
10 g Tryptone
10 g yeast extract
10 g K_2HPO_4
15 g agar
5 g lactose
1000 g distilled water

Adjust pH to 7.0.

Lascelles' medium (Ex. 27)
A: 3.8 sodium L-glutamate monohydrate
2.7 g DL-malic acid
0.5 g KH_2PO_4
1 mg nicotinic acid
1 mg thiamine-HCl
0.1 ml biotin (1 mg/10 ml water)
0.8 g $(NH_4)_2HPO_4$
900 ml distilled water
Adjust pH to 6.8 with 1 N NaOH; autoclave 10 minutes.
B: 0.2 g $MgSO_4 \cdot 7 H_2O$
40 mg $CaCl_2$
100 ml distilled water
Adjust pH to 6.8 with 1 N NaOH; autoclave 10 minutes.
Aseptically combine A and B.

Litmus milk See **Milk, litmus**

Lysine decarboxylase broth (Ex. 36)
5 g thiotone peptone
5 g beef extract
0.01 g brom cresol purple
0.005 g cresol red
0.5 g glucose
0.005 g pyridoxal
10 g L-lysine or 20 g DL-lysine
1000 ml distilled water
Adjust pH to 6.0.

M-17 broth (Ex. 50)
5 g polypeptone
2.5 g yeast extract
5 g glucose
19 g β-glycerolphosphate
5 g phytone peptone
5 g beef extract
0.5 g ascorbic acid
0.246 g $MgSO_4 \cdot 7 H_2O$
1000 ml distilled water
For *M-17 agar*, add 15 g agar per liter.

Manganese agar (sporulation agar) (Ex. 43, 46)
8 g nutrient broth
0.5 ml $MnCl_2$ solution (10 mg/ml solution)
20 g agar
1000 ml distilled water

Mannitol media (for *Azotobacter*) (Ex. 69)
15 g mannitol
0.5 g K_2HPO_4
0.2 g $MgSO_4$

0.1 g $CaSO_4$
0.2 g NaCl
5 g $CaCO_3$
1000 ml distilled water

Adjust pH to 8.3. For *mannitol agar*, add 15 g agar per liter.

Milk, litmus (Ex. 36)

Add sufficient azolitmin (2.5% aqueous solution) to fresh skim milk to give a lilac color. Autoclave 12 minutes at 15 psi and cool in water immediately following removal from the autoclave.

Minimal broth (Ex. 32)

A: 12 g KH_2PO_4
   12 g K_2HPO_4
   4 g NH_4Cl
   800 ml distilled water
   Adjust pH to 6.8–7.0.
B: 10 mg $FeSO_4 \cdot 7\ H_2O$
   40 g glucose
   100 mg $MgSO_4 \cdot 7\ H_2O$
   200 ml distilled water
   Adjust pH to 6.8–7.0.

Autoclave A and B separately. Combine aseptically.

Minimal Agar Per Liter (Ex. 51)

A: $(NH_4)_2\ SO_4$, 2.0 g
   K_2HPO_4, 14.0 g
   KH_2PO_4, 6.0 g
   Sodium Citrate, 1.0 g
   $MgSO_4\ 7\ H_2O$, 0.2 g
   Distilled water, 500 ml
B: Nobel agar, 15.0 g
   Distilled water, 500 ml

Autoclave separately, cool, to 50°C.
Mix A and B together.
Add the following sterile solutions aseptically:
50% glucose, 10 ml
10^{-3} M $MnCl_2$, 10 ml
Thiamine (0.337 μM), 5 ml

Required for recipient:
L-tryptophane, Stock 200 (0.41 μM), 5 ml
L-leucine, (0.79 μM), 5 ml
L-proline (4.6 μM), 5 ml
Adenine 0.02 mg/ml final concentration

Optional:
Streptomycin
Use streptomycin sulfate stored at 4°C, final concentration 200 μg/ml (100 mg per 500 ml of medium)

MR–VP broth (Ex. 35)

5 g glucose
5 g proteose peptone

5 g K_2HPO_4
Do not alter pH.

Neurospora minimal medium (Ex. 26)

2.5 g trisodium citrate
5 g KH_2PO_4
2 g NH_4NO_3
0.2 g $MgSO_4 \cdot 7\ H_2O$
0.1 g $CaCl_2 \cdot 2\ H_2O$
1000 ml distilled water
0.1 ml trace element solution (see below)
1 ml biotin solution (5 μg per ml)
20 g sucrose
20 g agar

Do not heat. Shake or stir medium to make certain that everything is dissolved before adding the next component.

Trace element solution: 5 g citric acid
5 g zinc sulfate
1 g $Fe(NH_4)_2\ (SO_4)_2 \cdot 6H_2O$
0.25 g $CuSO_4 \cdot 5H_2O$
0.05 g $MnSO_4 \cdot H_2O$
0.05 g boric acid
0.05 g $Na_2MoO_4 \cdot 2H_2O$
95 ml distilled water

Niacin assay broth (DIFCO B322) (Ex. 25)

12 g vitamin-free casein hydrolysate
40 g glucose
20 g sodium acetate
1 g K_2HPO_4
1 g KH_2PO_4
0.4 g $MgSO_4$
0.02 g NaCl
0.02 g $FeSO_4$
0.02 g $MnSO_4$
0.4 g L-cysteine
0.2 g DL-tryptophan
0.02 g adenine sulfate
0.2 mg thiamine HCl
0.2 mg calcium pantothenate
0.4 mg pyridoxine HCl
0.4 mg riboflavin
0.1 mg *p*-aminobenzoic acid
0.8 ug biotin

Niacin solution (Ex. 25)

Use a series: 0.0, 0.025, 0.05, 0.075, 0.1, 0.2, 0.3, 0.5 μg per tube of medium. These solutions may be prepared by first dissolving 0.1 g of niacin in 10 ml distilled water, fiving a stock solution of 10 mg/ml. To prepare the various concentrations, dilute the stock solution by adding 0.1 ml to 999 ml of distilled water, giving a 1 μg/ml solution.

Following Table A-2, make up 500 ml quantities of the media at each concentration of niacin. Dispense 10 ml of each concentration into a clear, unscratched, 18 mm test tube and autoclave for 12 minutes. All dilutions must be sterilized in the autoclave at the same time. The amounts given in the chart will make 50 tubes of each niacin concentration.

Table A-2   Niacin Solutions for Niacin Assay

μg Niacin/ Tube	Medium (ml)	Distilled Water (ml)	Niacin (1.0 μg/ml)
0	250	250	——
0.025	250	237	13
0.05	250	225	25
0.075	250	212.5	37.5
0.1	250	200	50
0.2	250	150	100
0.3	250	100	150
0.5	250	——	250

Nitrate broth (Ex. 69)
    5 g KNO_3 or $NaNO_3$
    1000 ml nutrient broth or sugar base broth

Nitrate-formation medium (Ex. 69)
    1 g $NaNO_2$
    1 g K_2HPO_4
    0.3 g $MgSO_4$
    1 g Na_2CO_3
    0.5 g NaCl
    0.4 g $FeSO_4$
    1000 ml distilled water
    This medium does not need to be sterilized.

Nitrate-reduction medium (Ex. 44)
    10 g casitone
    3 g yeast extract
    2 g potassium nitrate
    1000 ml distilled water
    Adjust pH to 7.1.

Nitrite-formation medium (Ex. 69)
    2 g $(NH_4)_2SO_4$
    1 g K_2HPO_4
    0.5 g $MgSO_4$
    0.4 g $FeSO_4$
    5–10 g $CaCO_3$
    1000 ml distilled water
    This medium does not have to be sterilized.

Novobiocin-penicillin-cyclohexamide agar
(Ex. 44)
    45 mg novobiocin
    44.9 mg penicillin G
    75 mg cyclohexamide (actidione)
    3 ml 95% ethanol
    Mix antibiotics in the ethanol. Dilute with 50 ml of sterile distilled water. Add to 1 liter of sterilized, tempered *Pseudomonas* agar F.

Nutrient media (broth)
    3 g beef extract
    5 g tryptone
    1000 ml distilled water
    Adjust pH to 7.0.
    For *nutrient agar*, add 15 g agar per liter.
    For *soft nutrient agar*, add 7 g agar per liter.
    For *nutrient agar plus NaCl*, add 15 g agar, 15 g NaCl per liter.
    For *nutrient gelatin*, add 40 g gelatin per liter.

Ornithine decarboxylase broth (Ex.36)
    5 g thiotone peptone
    5 g beef extract
    0.001 g brom cresol purple
    0.005 g cresol red
    0.5 g glucose
    0.005 g pyridoxal
    10 g L-ornithine or 20 g DL-ornithine
    1000 ml distilled water
    Adjust final pH to 6.8.

Oxidation-fermentation agar (Ex. 35, 44)
    2 g tryptone
    5 g NaCl
    3 g K_2HPO_4
    2 g agar
    0.08 g brom thymol blue
    1 g glucose
    1000 ml distilled water
    Autoclave for 10 minutes.

Phenol-red carbohydrate medium (Ex. 18, 35)
    10 g proteose peptone
    5 g NaCl
    0.018 g phenol red
    1000 ml distilled water
    5 g of desired carbohydrate
    Adjust pH to 7.4. For double strength medium, use 500 ml distilled water.

Phenylalanine agar (Ex. 36)
    2 g DL-phenylalanine
    3 g yeast extract
    5 g NaCl
    1 g Na_2HPO_4

12 g agar
Adjust pH to 7.3.

Phosphate-free Tris medium (for alkaline phosphatase assay) (Ex. 48)

Solution A (10×): 120 g Tris
20 g KCl
20 g NH_4Cl
1000 ml glass distilled water

Add chemicals to 900 ml water. Adjust pH to 7.6. This will take 60–68 ml concentrated HCl. Bring volume up to 1 liter.

Solution B (100×): 0.01 g $NaNPO_4$
1000 ml glass distilled water

Solution will be added to media by students.

Solution C (100×): 1.5 ml concentrated HCl
5 g $MgCl_2 \cdot 6 H_2O$
5 g Na_2SO_4
0.029 ferous ammonium sulfate
100 ml glass distilled water

Final pH should be near 2.

Glucose solution: 20 g glucose
100 ml glass distilled water

Final medium: 10 ml solution A
1 ml solution C
1 ml glucose solution
90 ml glass distilled water

Dispense in new dilution bottles. Keep glassware as phosphate free as possible. Autoclave phosphate separately from the remainder of the media. This medium must not be sterilized in steam that contains an antifungal agent.

Plate count agar (Ex. 59)

5 g tryptone
2.5 g yeast extract
1g glucose
15 g agar
1000 ml distilled water.

Potato dextrose agar (Ex. 66)

infusion from 200 g potatoes in 1000 ml water
20 g dextrose
15 g agar

Pseudomonas agar F (Ex. 44)

20 g proteose peptone
10 g maltose
1.5 g K_2HPO_4
0.73 g $MgSO_4$
10 ml glycerol
15 g agar
1000 ml distilled water

This medium is available commercially as Flo agar.

Pseudomonas agar P (Ex. 44)

20 g peptone
2 g DL-alanine
10 g sodium citrate
8.6 g potassium sulfate
1.4 g KCl
1.4 g $MgSO_4$
10 ml glycerol
15 g agar
1000 ml distilled water

This medium is available commercially as Tech agar.

Sabouraud dextrose agar (Ex. 65, 67)

10 g neopeptone
40 g dextrose
15 g agar
1000 ml distilled water
Adjust pH to 5.6.

Selenite cystine broth (Ex. 61)

5 g polypeptone peptone
4 g lactose
10 g Na_2HPO_4
4 g $NaHSeO_4$
0.01 g L-cystine
1000 ml distilled water
Adjust pH to 7.0.

Simmon's citrate agar (Ex. 35)

0.2 g $MgSO_4$
1 g $(NH_4)H_2PO_4$
1 g K_2HPO_4
2 g sodium citrate
5 g NaCl
15 g agar
0.08 g brom thymol blue
1000 ml distilled water

Dispense into test tubes and place in a slanted position.

Skim-milk agar See Casein agar

Synder test agar (Ex. 75)

20 g tryptone
20 g dextrose
5 g NaCl
20 g agar
0.02 g brom cresol green
1000 ml distilled water
Adjust pH to 4.8.

Starch agar (Ex. 35, 43)

10 g tryptone
10 g yeast extract
5 g K_2HPO_4

3 g soluble starch
15 g agar
1000 ml distilled water

Add starch to a small amount of water. Heat with constant stirring until the starch dissolves. Bring the solution to boiling and immediately add to the other ingredients.

Sulfide indole motility (SIM) agar (Ex. 36)
3 g beef extract
30 g peptone
0.2 g peptonized iron
0.025 sodium thiosulfate
3 g agar
1000 ml distilled water

Tech agar   See *Pseudomonas* **agar P**

Thioglycollate media (Ex. 29)
15 g pancreatic digest of casein
5 g L-cystine
5 g glucose
5 g yeast extract
2.5 g NaCl
0.5 g sodium thioglycollate
0.7 g agar
1000 ml distilled water
Adjust pH to 7.1.
For *thioglycollate agar*, add 15 g agar per liter.

Top agar
0.7% agar in medium

Trypticase-soy broth (Ex. 31)
17 g trypticase
3 g phytone
5 g NaCl
2.5 g K_2HPO_4
2.5 g glucose

1000 ml distilled water
Adjust ph to 7.3.
For *trypticase-soy agar*, add 15 g agar per liter.
For *trypticase-soy top agar*, add 7 g agar per liter.

Tryptone broth (Ex. 36)
10 g tryptone
1000 ml distilled water

TY agar (Ex. 68) See **Yeast extract tryptone medium**

TYG broth (Ex. 51, 53)
5 g tryptone
5 g yeast extract
1 g glucose
1 g K_2HPO_4
1000 ml tap water
Adjust pH to 7.0.
For TYG agar, add 15 g agar per liter.

Urea broth (Ex. 37)
10 g urea
1 g peptone
5 g NaCl
1 g glucose
2 g KH_2PO_4
0.012 g phenol red
1000 ml distilled water
Adjust pH to 6.8 – 6.9. Filter sterilize and aseptically dispense to sterile tubes.

Yeast extract tryptone medium (Ex. 28, 38, 40)
10 g tryptone
5 g yeast extract
5 g K_2HPO_4
1 g glucose
1000 ml distilled water
For *yeast extract tryptone agar*, add 15 g agar per liter.

INDEX